ICT and New Media Bridging Technology and Communication

ICT and New Media Bridging Technology and Communication

Ashutosh Sharma
Department of Agricultural Extension,
Rani Lakshmi Bai Central Agricultural University, Jhansi, U.P. India

Deepak Shrivastava
Department of Agriculture Extension,
BTC CARS, Bilaspur, IGKV, Raipur. C.G. India

CRC Press
Taylor & Francis Group
Boca Raton London New York

CRC Press is an imprint of the
Taylor & Francis Group, an **informa** business

-EPH-

Elite Publishing House

First published 2026
by CRC Press
4 Park Square, Milton Park, Abingdon, Oxon, OX14 4RN

and by CRC Press
2385 NW Executive Center Drive, Suite 320, Boca Raton FL 33431

British Library Cataloguing-in-Publication Data
A catalogue record for this book is available from the British Library

Print edition not for sale in India.

ISBN13: 9781041247302 (hbk)
ISBN13: 9781041264675 (pbk)
ISBN13: 9781003750444 (ebk)

DOI: 10.4324/9781003750444

Typeset in Adobe Caslon Pro
by Elite Publishing House, Delhi

-EPH-

Contents

Preface

In the contemporary era, Information and Communication Technology (ICT) and New Media are revolutionizing the way we live, learn, and communicate. These transformative fields not only reshape industries but also redefine societal norms, enabling individuals and communities to connect in unprecedented ways. The convergence of technology, communication, and creativity forms the backbone of a world that thrives on innovation and adaptability.

This book, "ICT and New Media: Bridging Technology and Communication," has been carefully designed to address the multifaceted dimensions of ICT and New Media. It reflects my deep commitment to fostering an understanding of these fields and their profound influence on various aspects of modern life. The book is a structured blend of foundational principles, emerging trends, and practical applications aimed at empowering readers with both theoretical insights and hands-on expertise.

The inspiration for this work arises from the pressing need to provide a resource that is not only informative but also forward-looking. As technology advances at an extraordinary pace, the ability to comprehend and adapt to these changes is critical. This book seeks to bridge the gap between knowledge and application, enabling readers to engage with ICT and New Media in meaningful and innovative ways. Organized into four distinct units, the book takes readers on a journey through the essential elements of ICT and New Media, from foundational concepts to cutting-edge technologies. Each unit has been meticulously curated to provide a comprehensive understanding of the subject while maintaining relevance to the demands of today's technological landscape.

We hope this book serves as a valuable resource for students, educators, professionals, and researchers. May it inspire you to explore, innovate, and contribute to the ever-expanding world of ICT and New Media

Authors

About the Authors

Dr. Ashutosh Sharma is working as an Assistant Professor and is actively involved in agricultural work at Rani Lakshmi Bai Central Agricultural University, Jhansi. He did B.Sc. (Ag.) and M.Sc. (Ag.) in agricultural extension from Indira Gandhi Krishi Vishwavidyalaya, Raipur, Chhattisgarh and Ph.D. from Jawahar Lal Nehru Krishi Vishwavidyalaya, Jabalpur, Madhya Pradesh. He has published more than 15 research papers in national and International journals. He is also credited with three books and nine book chapters in his curriculum. He has also received various scientific honors in the field of agricultural extension including Young Scientist, scientist associate award, best young teacher award best agricultural extension worker award, etc.

Dr. Deepak Shrivastava, Assistant Professor at BTC CARS, Bilaspur under IGKV Raipur, specializes in Agricultural Extension with over 20 years of research and teaching experience. He holds a B.Sc. and M.Sc. from JNKVV, Jabalpur, Madhya Pradesh, and a Ph.D. from Guru Ghasidas Central University, Bilaspur, Chhattisgarh. Dr. Shrivastava has contributed significantly to agricultural research with 10 published research papers, attended 20 conferences, and authored 10 Hindi articles. He has actively participated in 10 training programs and organized 20 training sessions, reflecting his dedication to capacity building. His extensive work includes KVK activities, farm research, and 15 radio talks aimed at disseminating agricultural knowledge. Dr. Shrivastava's expertise bridges academic research and practical applications, making valuable contributions to agricultural development. His work in underutilized crops, combined with his passion for extension services, positions him as a key figure in advancing sustainable agricultural practices.

Chapter 1

Role and Components of ICT in Community Education

Introduction

Information Communication Technology (ICT) stands as a transformative force in modern society, shaping how individuals, organizations, and governments interact and operate. As a field that combines computing and communication technologies, ICT provides the tools and frameworks necessary for processing, managing, and disseminating information efficiently. Its widespread adoption across industries has made ICT indispensable in addressing challenges, fostering innovation, and enabling socio-economic development.

The scope of ICT is vast, encompassing a wide range of technologies, applications, and services. At its core are several fundamental components, such as hardware, software, networks, and data, which work together to create systems that support information exchange. These components form the foundation for advanced ICT applications, including IT-enabled services like call centers, helpdesks, and data warehouses. These services not only enhance organizational efficiency but also create significant opportunities for economic and social development. ICT's role in community education is particularly noteworthy. By providing access to digital learning tools, e-resources, and collaborative platforms, ICT has revolutionized education delivery, making learning more accessible, interactive, and inclusive. This capability is critical in bridging the digital divide and empowering underserved communities with knowledge and skills essential for socio-economic advancement.

Governments worldwide have recognized the transformative potential of ICT and have formulated policies to support its integration across sectors. These policies aim to enhance digital infrastructure, promote ICT literacy, and ensure equitable access to technology. By fostering an environment conducive to innovation and digital inclusion, governments aim to leverage ICT as a driver of development. Despite its advancements, ICT continues to evolve, presenting emerging research opportunities and challenges. From addressing cybersecurity threats to developing sustainable ICT solutions, researchers are exploring new frontiers to maximize the potential of ICT while mitigating its risks. The integration of emerging technologies such as artificial intelligence (AI), machine learning (ML), and blockchain further expands the scope of ICT, opening new avenues for innovation.

A. Components of Information Communication Technology (ICT)

Information Communication Technology (ICT) is a broad term encompassing the technologies,

DOI: 10.4324/9781003750444-1

tools, and systems used to manage, process, and communicate information. The components of ICT form the building blocks that support its applications and functionalities across various industries. Understanding these components is essential for grasping the intricacies of how ICT operates and contributes to societal progress.

1. HARDWARE

Hardware represents the tangible, physical components of Information Communication Technology (ICT), forming the backbone upon which digital systems are built and operated. These devices and equipment facilitate the input, processing, storage, output, and communication of information, enabling a wide array of ICT applications. From personal computing devices to large-scale data centers, hardware serves as the interface between the digital and physical realms, transforming raw data into actionable insights and enabling connectivity in an increasingly interconnected world.

Key Categories of Hardware

a. Input Devices

Input devices are indispensable components of any Information and Communication Technology (ICT) system, acting as the primary interface between users and machines. They facilitate the entry of data into digital systems, serving as the critical first step in the transformation of raw input—whether it be text, sound, images, or environmental signals—into digital information for processing and analysis. Input devices bridge the gap between human intent and computational processes, ensuring seamless communication and interaction within the digital ecosystem. The evolution of input devices has paralleled the advancements in ICT, resulting in increasingly sophisticated tools that cater to a diverse range of applications. From the traditional keyboard and mouse to modern sensors embedded in IoT devices, input technologies have expanded how users can interact with systems.

1. Keyboards and Mice

Keyboards and **mice** are the most ubiquitous and widely recognized input devices. Together, they form the cornerstone of user interaction in traditional computing environments.

» **Keyboards**: Keyboards are designed primarily for text input, featuring a grid of keys that represent letters, numbers, symbols, and special functions. The standard QWERTY keyboard layout is the most common, though alternative layouts such as AZERTY and DVORAK exist to cater to regional preferences and specialized tasks. Advanced keyboards now include features like:

- **Mechanical Switches** for tactile feedback.
- **Backlit Keys** for visibility in low-light conditions.
- **Programmable Keys** for custom shortcuts, particularly useful in gaming or professional workflows.

» **Mice**: Mice are pointing devices that allow users to navigate and interact with graphical interfaces. They typically include buttons for selection and scrolling mechanisms for vertical or horizontal movement.

Modern innovations in mice include:

- **Optical and Laser Sensors** for precision tracking.
- **Ergonomic Designs** to reduce strain during extended use.
- **Wireless Connectivity** for increased mobility.

Applications: Keyboards and mice remain the primary tools for tasks such as typing documents, navigating user interfaces, coding, and engaging in design or gaming activities.

2. Scanners

Scanners are devices that digitize physical objects, converting them into digital formats for storage, editing, or analysis. By capturing images, text, or patterns from tangible sources, scanners expand the scope of input data.

» **Document Scanners**: Convert physical documents into digital text or image files, often used in offices for digitizing records.

» **Barcode Scanners**: Read barcodes for inventory management, retail transactions, and logistics tracking.

» **Flatbed Scanners**: Offer high-resolution scanning for photographs and other delicate materials.

» **3D Scanners**: Capture the geometry of three-dimensional objects, enabling applications in design, manufacturing, and virtual reality.

Applications: Scanners are integral to workflows that require the conversion of physical data into digital repositories, such as archiving, graphic design, and retail operations.

3. Voice Recognition Devices

Voice recognition devices have revolutionized the way users interact with technology by enabling hands-free, intuitive communication. These devices convert spoken language into digital commands, text, or actions, leveraging advanced algorithms and artificial intelligence for high accuracy.

» **Virtual Assistants**: Tools like Amazon Alexa, Google Assistant, and Apple Siri interpret voice commands to perform tasks, answer queries, and control smart devices.

» **Speech-to-Text Systems**: Convert spoken words into written text, used for dictation, transcription, and accessibility services.

» **Command-Based Interfaces**: Found in applications like customer service systems or automotive controls, where specific commands trigger predefined actions.

Applications: Voice recognition devices are widely used in accessibility technologies, smart home systems, automated customer support, and productivity tools that prioritize convenience and efficiency.

4. Touchscreens and Styluses

Touchscreens and styluses have become essential input technologies in devices ranging from smartphones to advanced industrial machinery. They offer direct, intuitive interaction with digital interfaces, eliminating the need for intermediary devices like keyboards or mice.

- » **Touchscreens:** Touch-sensitive screens detect user input through contact, whether by finger, stylus, or multi-touch gestures. Capacitive touchscreens are the most common type, used in devices such as smartphones, tablets, and kiosks.
- » **Styluses:** A stylus is a pen-like device designed for precision input on touch-enabled devices. It is commonly used in applications that require detailed interaction, such as drawing, note-taking, and photo editing. Advanced styluses incorporate pressure sensitivity, tilt detection, and customizable buttons to enhance functionality.

Applications: Touchscreens and styluses are integral to industries like design, education, healthcare (e.g., touch interfaces in medical devices), and retail (e.g., self-checkout systems).

5. Sensors

Sensors represent a significant leap in input device technology, capturing environmental data and converting it into digital signals for analysis. They are at the core of IoT systems, enabling devices to sense and respond to their surroundings autonomously.

- » **Temperature Sensors**: Monitor environmental heat, used in HVAC systems, industrial processes, and wearable health devices.
- » **Motion Sensors**: Detect movement, used in security systems, gaming consoles, and augmented reality applications.
- » **Light Sensors**: Adjust screen brightness based on ambient light, found in smartphones and automotive systems.
- » **Proximity Sensors**: Detect nearby objects, used in smartphones for call management and in robotics for obstacle avoidance.
- » **Biometric Sensors**: Measure biological characteristics like fingerprints, heart rate, or facial features, enabling security features and health monitoring.

Applications: Sensors are critical in IoT ecosystems, automation systems, environmental monitoring, and personalized healthcare technologies.

Advancements and Emerging Trends in Input Devices

Input devices continue to evolve, incorporating new technologies to enhance functionality, accuracy, and user experience:

1. **Haptic Feedback**: Devices that provide tactile responses to user actions, creating a more immersive interaction.
2. **Gesture Recognition**: Systems that interpret hand or body movements as input commands, popularized by devices like Microsoft Kinect.
3. **Brain-Computer Interfaces (BCIs)**: Advanced systems that translate neural signals into commands, offering groundbreaking applications for accessibility and gaming.
4. **Augmented Reality (AR) and Virtual Reality (VR) Controllers**: Specialized input devices that enable users to interact with virtual environments.

5. **Multimodal Input**: Integration of multiple input methods (e.g., voice, touch, gesture) for seamless and adaptive user experiences.

b. Processing Units

Processing units are the central components of any Information and Communication Technology (ICT) system, functioning as the core computational entities that execute instructions, process data, and enable complex operations. These units form the heart of both general-purpose computing devices and specialized hardware systems, driving the performance and efficiency of modern technology. Processing units operate by interpreting and executing machine-level instructions derived from software applications. This capability allows them to perform logical, arithmetic, and control functions essential for a vast range of tasks, from simple computations to high-performance scientific simulations. Over time, advancements in processing technology have significantly enhanced the speed, efficiency, and versatility of these units, enabling breakthroughs in artificial intelligence, big data analytics, and real-time communication systems.

Key Processing Units

1. **Central Processing Units (CPUs)** The **Central Processing Unit (CPU)** is often referred to as the "brain" of a computer, as it manages and coordinates most of the system's operations. CPUs are designed to handle a wide array of general-purpose tasks, making them integral to virtually all computing devices, from personal computers to servers.

   » **Core Functions**:

   - Executing instructions from software applications.
   - Performing arithmetic and logical operations.
   - Managing data flow between memory, input/output devices, and other system components

   » **Key Features**:

   - **Cores**: Modern CPUs are multi-core, meaning they have multiple processing units within a single chip, enabling parallel processing for improved performance.
   - **Clock Speed**: Measured in gigahertz (GHz), clock speed determines how many cycles a CPU can execute per second, influencing its overall speed.
   - **Cache Memory**: On-chip memory that provides high-speed access to frequently used data, reducing latency and enhancing performance.

Applications: CPUs are essential in general-purpose computing, supporting tasks such as word processing, web browsing, data analysis, and software development. They are also used in embedded systems, powering devices like smartphones, household appliances, and automobiles.

2. **Graphics Processing Units (GPUs):** Originally developed for rendering high-quality graphics, **Graphics Processing Units (GPUs)** have evolved into powerful processors capable of performing highly parallel tasks. GPUs are particularly well-suited for applications requiring large-scale data processing and computations.

» **Core Functions**:

- Rendering 2D and 3D graphics for visual output.
- Performing parallel computations in applications like deep learning, scientific simulations, and video encoding.

» **Key Features**:

- **Parallel Architecture**: GPUs consist of thousands of smaller cores designed to handle multiple tasks simultaneously, making them ideal for computationally intensive workloads.
- **Specialized APIs**: Tools like CUDA (for NVIDIA GPUs) and OpenCL enable developers to harness GPU power for general-purpose computing tasks.

Applications: GPUs are widely used in gaming, video editing, and animation, as well as in emerging fields like artificial intelligence (AI), machine learning (ML), and cryptocurrency mining. In these contexts, their ability to process vast amounts of data in parallel significantly outperforms traditional CPUs.

3. **Application-Specific Integrated Circuits (ASICs): Application-Specific Integrated Circuits (ASICs)** are custom-designed hardware components tailored for specific tasks. Unlike general-purpose CPUs and GPUs, ASICs are optimized to execute particular operations with unmatched efficiency and performance.

   » **Core Functions**:

   - Performing specialized computations with minimal power consumption.
   - Executing tasks at higher speeds compared to general-purpose processors.

   » **Key Features**:

   - **Task-Specific Design**: ASICs are engineered to handle dedicated tasks, such as encoding video, processing network data, or mining cryptocurrencies.
   - **Energy Efficiency**: Because they are designed for specific functions, ASICs consume less power than CPUs and GPUs for the same task.

Applications: ASICs are extensively used in telecommunications (e.g., modems), consumer electronics (e.g., image processors in cameras), and blockchain technologies (e.g., Bitcoin mining rigs). They also play a significant role in industrial automation and aerospace applications, where precision and reliability are paramount.

4. **Field-Programmable Gate Arrays (FPGAs)**: **Field-Programmable Gate Arrays (FPGAs)** are versatile processing units that can be reconfigured after manufacturing to perform specific tasks. This flexibility makes them invaluable for prototyping, testing, and deploying custom hardware solutions.

   » **Core Functions**:

   - Enabling hardware customization for specific use cases.
   - Supporting rapid prototyping and iterative design processes.

» **Key Features**:

- **Reprogrammability**: FPGAs can be reconfigured to support evolving requirements or new standards.
- **Parallel Processing**: Like GPUs, FPGAs can execute multiple operations simultaneously, though their architecture is more adaptable.
- **High Performance**: They offer near-ASIC performance for specific applications, but without the need for costly custom manufacturing.

Applications: FPGAs are used in industries such as telecommunications (e.g., 5G base stations), automotive (e.g., autonomous driving systems), and finance (e.g., high-frequency trading platforms). Their adaptability also makes them valuable in research and development environments where flexibility is critical.

Advancements in Processing Units

The field of processing technology is continually advancing, driven by the demand for faster, more efficient, and specialized computing solutions. Key trends include:

1. **Multi-Core and Many-Core Architectures**: The shift from single-core to multi-core processors has enabled systems to handle multiple tasks simultaneously. Many-core architectures, such as those found in GPUs, extend this concept, providing thousands of cores for parallel processing.
2. **Heterogeneous Computing**: Modern systems increasingly combine CPUs, GPUs, and other specialized units (e.g., ASICs, FPGAs) to achieve optimal performance for a variety of tasks. This approach leverages the strengths of each processing unit for specific workloads.
3. **Energy Efficiency**: Advances in semiconductor manufacturing, such as the transition to smaller process nodes (e.g., 7nm and 5nm), have improved the energy efficiency of processing units. Low-power processors are essential for mobile devices, IoT applications, and data centers.
4. **Quantum Processing Units (QPUs)**: Quantum computing represents the frontier of processing technology, offering the potential to solve problems that are intractable for classical computers. Though still in development, QPUs are poised to revolutionize fields such as cryptography, optimization, and molecular modeling.
5. **AI Accelerators**: The growing demand for AI and ML applications has spurred the development of specialized processing units, such as Tensor Processing Units (TPUs) and neural processing units (NPUs). These accelerators are optimized for tasks like training deep learning models and performing inference.

c. Storage Devices

Storage devices are critical components of Information and Communication Technology (ICT) systems, responsible for retaining digital data for both short-term operations and long-term archival purposes. These devices enable users and systems to store, retrieve, and manage information efficiently, forming the backbone of modern digital workflows. The evolution of storage technologies has greatly influenced the speed, reliability, and scalability of ICT systems, paving the way for innovations in data management and access. Storage devices are broadly categorized based on their functionality, technology, and form factor. From traditional mechanical drives to cutting-edge cloud-based solutions, these devices have undergone significant advancements to meet the growing demands of data-centric applications.

Key Categories of Storage Devices

1. **Hard Disk Drives (HDDs): Hard Disk Drives (HDDs)** are among the oldest and most widely used storage technologies, utilizing magnetic storage principles to retain data. HDDs consist of spinning platters coated with a magnetic material and a read/write head that moves across the surface to access or modify data.

   » **Features:**

     - **Capacity**: HDDs are known for their large storage capacities, often ranging from a few hundred gigabytes (GB) to multiple terabytes (TB).
     - **Cost-Effectiveness**: HDDs provide high storage capacities at relatively low costs, making them ideal for bulk data storage.
     - **Durability**: While HDDs are reliable, their mechanical components are susceptible to wear and physical damage over time.

   » **Applications**: HDDs are commonly used in desktop computers, servers, and backup systems where cost-effective, high-capacity storage is required.

   » **Limitations:**

     - Slower data access speeds compared to modern alternatives.
     - Increased vulnerability to mechanical failure due to moving parts.

2. **Solid-State Drives (SSDs): Solid-State Drives (SSDs)** represent a significant leap forward in storage technology, offering faster and more reliable data access compared to HDDs. Unlike HDDs, SSDs use flash memory to store data, eliminating the need for mechanical components.

   » **Features:**

     - **Speed**: SSDs provide faster read and write speeds, resulting in quicker boot times, application launches, and file transfers.
     - **Durability**: With no moving parts, SSDs are less prone to physical damage and mechanical failure.
     - **Energy Efficiency**: SSDs consume less power than HDDs, making them ideal for mobile and energy-conscious applications.

   » **Applications**: SSDs are widely used in laptops, gaming systems, high-performance computing, and data-intensive applications such as video editing and database management.

   » **Limitations:**

     - Higher cost per GB compared to HDDs.
     - Limited write cycles, though advancements in technology have significantly mitigated this issue.

3. **Optical Storage Devices**: **Optical Storage Devices** use laser technology to read and write data on discs such as CDs, DVDs, and Blu-rays. These devices offer a cost-effective solution for distributing and archiving data.
   - » **Features**:
     - **Portability**: Optical discs are lightweight and easy to transport.
     - **Longevity**: Optical storage offers good data retention over long periods if stored under optimal conditions.
     - **Capacity**: Varies by format, with CDs offering up to 700 MB, DVDs up to 4.7 GB (single-layer), and Blu-rays up to 50 GB (dual-layer).
   - » **Applications**: Optical devices are commonly used for media distribution (movies, music, software) and archival purposes, particularly in scenarios where long-term storage is prioritized over frequent access.
   - » **Limitations**:
     - Limited storage capacity compared to HDDs and SSDs.
     - Slower read/write speeds.
     - Vulnerability to physical damage, such as scratches.
4. **Removable Storage**: **Removable Storage Devices** are portable storage solutions that allow users to transfer data easily between systems. These devices are compact, lightweight, and convenient, making them popular for personal and professional use.
   - » **Types**:
     - **USB Drives**: Small, portable devices offering storage capacities ranging from a few GBs to several TBs.
     - **Memory Cards**: Used in cameras, smartphones, and other portable devices for storing photos, videos, and other data.
     - **External Hard Drives**: Portable versions of HDDs or SSDs, often used for data backups and large file transfers.
   - » **Applications**: Removable storage is widely used for data sharing, backups, and temporary storage needs, particularly in mobile environments.
   - » **Limitations**:
     - Susceptibility to loss or theft due to small size.
     - Limited durability and lifespan, especially for frequently used devices.
5. **Cloud-Based Storage Solutions**: **Cloud-Based Storage** represents the forefront of modern storage technology, offering virtualized storage services accessible over the internet. These solutions eliminate the need for physical storage devices, providing unparalleled scalability and flexibility.

- **Features**:
  - **Accessibility**: Users can access data from any internet-enabled device, promoting collaboration and mobility.
  - **Scalability**: Storage capacity can be expanded or contracted as needed, ensuring cost efficiency.
  - **Redundancy**: Cloud providers ensure data redundancy, protecting against data loss due to hardware failures.
- **Applications**: Cloud storage is extensively used in personal and professional environments for file sharing, backups, and collaboration. It supports critical business applications such as enterprise resource planning (ERP) and customer relationship management (CRM) systems.
- **Limitations**:
  - Dependent on reliable internet connectivity.
  - Potential security and privacy concerns if proper safeguards are not implemented.

Advancements and Trends in Storage Devices

The landscape of storage technology continues to evolve, driven by the exponential growth of data and the demand for faster, more efficient systems. Key trends include:

1. **Non-Volatile Memory Express (NVMe) Drives**: Offering faster data transfer rates than traditional SSDs, NVMe drives are becoming the standard in high-performance storage.
2. **Hybrid Storage Solutions**: Combining the speed of SSDs with the capacity of HDDs to optimize performance and cost-efficiency.
3. **Storage Class Memory (SCM)**: Bridging the gap between DRAM and NAND flash, SCM technologies such as Intel Optane provide low-latency, high-speed storage options.
4. **DNA Data Storage**: A cutting-edge technology that encodes data in synthetic DNA strands, promising unprecedented storage densities.
5. **Decentralized Storage**: Systems like IPFS (InterPlanetary File System) and blockchain-based solutions promote distributed storage models, enhancing resilience and security.

d. Output Devices

Output devices are essential components of Information and Communication Technology (ICT) systems, responsible for presenting the results of data processing to users in a form that is understandable and actionable. These devices act as the interface between digital systems and human perception, enabling individuals to interact with and interpret the information produced by computers and other digital technologies. The importance of output devices cannot be overstated, as they bridge the gap between raw computational outputs and human-readable formats. Whether it is displaying graphical content on a screen, producing hard copies of documents, or playing audio, output devices ensure that the digital realm communicates effectively with its users. With advancements in technology, these devices have become increasingly sophisticated, offering higher precision, clarity, and versatility.

Key Categories of Output Devices

1. **Monitors**: Monitors are the most common and widely used output devices, providing a visual representation of processed information. They range from basic displays to advanced screens with high-definition capabilities.
   - **Features**:
     - **Display Technology**: Includes Liquid Crystal Display (LCD), Light Emitting Diode (LED), Organic LED (OLED), and Quantum Dot displays.
     - **Resolution**: Determines the clarity of the image, ranging from standard HD to ultra-high-definition (4K and 8K).
     - **Refresh Rate**: Indicates how often the image is refreshed per second, critical for gaming and motion-heavy applications.
   - **Applications**: Monitors are used in various environments, including personal computing, gaming, graphic design, video editing, and data visualization. Advanced monitors are also employed in fields like medical imaging and simulation.
   - **Emerging Trends**:
     - Curved and ultrawide monitors for immersive experiences.
     - Touchscreen monitors that combine input and output functionalities.
     - Energy-efficient designs that reduce power consumption.
2. **Printers**: Printers convert digital information into tangible, physical forms, allowing users to produce hard copies of documents, images, or designs. They are indispensable in environments where physical documentation or presentation is required.
   - **Types of Printers**:
     - **Inkjet Printers**: Ideal for high-quality image printing, commonly used in homes and small offices.
     - **Laser Printers**: Known for speed and efficiency, preferred in office environments for bulk printing.
     - **3D Printers**: Revolutionizing manufacturing by creating three-dimensional objects from digital models.
     - **Dot Matrix Printers**: Still used in niche areas like industrial settings due to their ability to print on multi-layer forms.
   - **Applications**: Printers are widely used in education, business, publishing, and design industries. 3D printers, in particular, have opened up new possibilities in prototyping, medicine (e.g., prosthetics), and even construction.
   - **Advancements**:
     - Eco-friendly printers with reduced energy consumption and recyclable cartridges.
     - Wireless printing technologies for seamless operation across devices.

3. **Speakers and Headphones**: These output devices convert digital audio signals into sound, enabling users to listen to music, participate in virtual meetings, or consume other audio content.
   » **Features**:
      - **Speakers**: Range from simple, portable Bluetooth speakers to sophisticated surround sound systems for immersive audio experiences.
      - **Headphones**: Include over-ear, on-ear, and in-ear designs, with features like noise cancellation and wireless connectivity.
      - **Microphone Integration**: Many headphones now include built-in microphones for two-way communication.

   » **Applications**:
      - Media playback for entertainment and education.
      - Communication in virtual meetings and gaming.
      - Accessibility tools for visually impaired users, allowing them to interact with auditory feedback systems.

   » **Emerging Trends**:
      - Smart speakers powered by voice assistants like Alexa and Google Assistant.
      - High-fidelity (Hi-Fi) audio systems for professional music production and audiophiles.

4. **Projectors**: Projectors are devices that display visual content on large surfaces such as screens or walls. They are widely used in educational, corporate, and entertainment settings.
   » **Types of Projectors**:
      - **LCD Projectors**: Affordable and commonly used for presentations.
      - **DLP (Digital Light Processing) Projectors**: Offer sharper images and are suitable for cinematic applications.
      - **Laser Projectors**: High-end projectors known for brightness and longevity.

   » **Applications**:
      - Classrooms for interactive lessons and e-learning.
      - Boardrooms for business presentations and collaborative work.
      - Theaters for large-scale entertainment.

   » **Innovations**:
      - Portable projectors for on-the-go presentations.
      - Ultra-short-throw projectors that can display large images from a short distance.

5. **Braille Displays**: Braille displays are specialized output devices designed to assist visually impaired users by converting digital text into tactile braille characters. These devices provide accessibility to digital information, enabling greater inclusivity in ICT.

» **Features**:
  - Rows of braille cells that dynamically change based on the digital text being read.
  - Integration with screen readers to interpret on-screen content.
  - Portable and lightweight designs for easy mobility.

» **Applications**: Braille displays are used in education, employment, and personal computing by individuals with visual impairments. They facilitate access to books, emails, websites, and more.

» **Challenges**:
  - High cost due to the complexity of manufacturing.
  - Limited availability in developing regions.

Emerging Trends and Technologies in Output Devices

The field of output devices is witnessing rapid innovation, driven by advancements in display technology, audio systems, and accessibility tools. Some noteworthy trends include:

1. **Augmented Reality (AR) and Virtual Reality (VR) Displays**:
   - AR/VR devices provide immersive experiences by overlaying digital elements on the real world or creating entirely virtual environments.
   - Widely used in gaming, education, training, and simulations.
2. **Holographic Displays:**
   - These advanced displays project 3D images into space, offering an interactive and futuristic way to visualize data.
   - Applications include medical imaging, product design, and entertainment.
3. **Eco-Friendly Output Devices**:
   - Focus on reducing environmental impact with energy-efficient displays and recyclable materials.
   - Solar-powered devices are emerging as sustainable alternatives.
4. **Multi-Sensory Output Devices**:
   - Devices that combine visual, auditory, and tactile outputs to create richer user experiences.
   - Examples include haptic feedback in gaming controllers and multisensory training simulators.

e. Networking Hardware

Networking hardware forms the structural backbone of modern communication systems, enabling devices to interconnect, exchange data, and share resources. This critical infrastructure supports the seamless operation of networks in homes, businesses, and global enterprises. Networking hardware encompasses a variety of components, each serving specific roles in creating, maintaining, and optimizing networks for efficiency, security, and scalability. With the rapid growth of connected devices, fueled by the Internet of Things (IoT) and the increasing reliance on cloud computing, the demand for robust and sophisticated networking hardware has grown exponentially. These components are not only responsible for facilitating communication but also for ensuring that networks are secure, fast, and capable of handling the ever-growing volume of data traffic.

Core Networking Hardware Components

1. **Routers**: Routers are integral to networking infrastructure, directing data packets between different networks and ensuring efficient communication pathways. They function as traffic managers, determining the optimal route for data to travel from a source to its destination.

   » **Key Functions**:

   - **Routing Protocols**: Routers use protocols such as OSPF (Open Shortest Path First) and BGP (Border Gateway Protocol) to decide the best routes for data transmission.
   - **Network Address Translation (NAT)**: This feature enables multiple devices on a local network to share a single public IP address, conserving IP resources and enhancing security.
   - **Firewall Integration**: Many routers incorporate basic firewall capabilities to protect networks from unauthorized access.

   » **Applications**:

   - **Home Networks**: Routers connect devices like smartphones, laptops, and smart appliances to the internet.
   - **Business Environments**: High-capacity routers manage complex traffic flows and ensure reliable connectivity for enterprise applications.
   - **Cloud Services**: Routers are critical in connecting users to cloud platforms and ensuring data flow between cloud regions.

   » **Advancements**:

   - **Wi-Fi 6 Routers**: Offer higher speeds, improved device handling, and reduced latency, ideal for IoT-heavy environments.
   - **Mesh Networking**: Multi-router systems that eliminate dead zones and provide consistent coverage across large areas.
   - **SD-WAN Routers**: Enable software-defined wide-area networks, offering dynamic path selection and improved network performance.

   » **Challenges**:

   - **Configuration Complexity**: Setting up advanced routing features requires technical expertise.
   - **Security Threats**: Improperly configured routers can become entry points for cyberattacks.

2. **Switches**: Switches are essential devices in Local Area Networks (LANs), enabling direct communication between connected devices. They operate at the data link layer (Layer 2) or network layer (Layer 3) of the OSI model, ensuring efficient intra-network communication.

   » **Key Features**:

   - **Packet Switching**: Switches forward data packets only to the intended recipient device, reducing unnecessary network traffic.
   - **VLAN Support**: Virtual LANs allow network segmentation, improving security and performance.
   - **PoE (Power over Ethernet)**: Combines data and power delivery over the same cable, simplifying device installation.

   » **Applications**:

   - **Office Environments**: Connect computers, printers, and servers for centralized resource sharing.
   - **Data Centers**: High-speed switches form the backbone of server communication and data flow.
   - **Industrial Automation**: Ruggedized switches enable communication between sensors, controllers, and other devices in harsh environments.

   » **Advancements**:

   - **10/40/100 Gigabit Switches**: Designed for high-bandwidth applications like video streaming and AI workloads.
   - **Cloud-Managed Switches**: Allow remote configuration and monitoring, reducing operational overhead.

   » **Challenges**:

   - **Cost**: High-performance switches can be expensive, especially for small businesses.
   - **Maintenance**: Regular updates and monitoring are necessary to ensure optimal performance and security.

3. **Modems**: Modems are critical for enabling internet connectivity by converting digital signals from devices into formats suitable for transmission over communication lines, such as telephone or cable systems.

   » **Types of Modems**:

   - **DSL Modems**: Utilize telephone lines for broadband internet.
   - **Cable Modems**: Connect to internet service providers via coaxial cables.
   - **Fiber Optic Modems**: Offer ultra-fast internet using optical fibers.

   » **Applications**:

   - Residential setups for home internet access.

- Enterprise networks requiring stable and high-speed internet connectivity.
- Rural and remote areas relying on satellite modems for connectivity.

» **Advancements**:

- **Integrated Modem-Routers**: Combine modem and router functions for simplified setups.
- **5G Modems**: Deliver ultra-fast mobile internet for on-the-go connectivity.
- **IoT-Specific Modems**: Optimized for low-power, long-range communication in IoT applications.

» **Challenges**:

- **Compatibility**: Ensuring modems work with specific ISP technologies can be a barrier.
- **Bandwidth Limitations**: Older modems may struggle with high-speed internet demands.

4. **Network Interface Cards (NICs)**: NICs are hardware components embedded in or added to devices to enable network connectivity. They serve as the physical interface between a device and a network.

» **Features**:

- **Wired NICs**: Use Ethernet cables for stable and high-speed connections.
- **Wireless NICs**: Provide Wi-Fi connectivity, supporting mobility and convenience.
- **Advanced NICs**: Support features like RDMA (Remote Direct Memory Access) for reduced latency in data-intensive applications.

» **Applications**:

- Computers, laptops, and servers for network access.
- IoT devices for integrating with wireless or wired networks.
- High-performance systems like gaming PCs and data servers.

» **Advancements**:

- **Multi-Gigabit NICs**: Capable of handling data transfer rates exceeding 1 Gbps.
- **Virtual NICs**: Enable network connectivity for virtual machines in cloud environments.

» **Challenges**:

- **Energy Efficiency**: NICs must balance performance with power consumption, especially in battery-powered devices.
- **Security**: NICs can be vulnerable to hardware-based attacks.

5. **Access Points**: Access points extend wireless connectivity in a network, enabling mobile devices to connect to the internet or other network resources. They act as wireless hubs, bridging wired and wireless segments of a network.

» **Features:**
  - **Dual-Band Support**: Operates on 2.4 GHz and 5 GHz bands to optimize performance.
  - **Centralized Management**: Large deployments can be managed from a single interface.
  - **Advanced Security**: Support for WPA3 encryption to secure wireless communication.

» **Applications:**
  - Homes and small businesses for Wi-Fi access.
  - Large enterprises for seamless connectivity across sprawling campuses.
  - Public spaces like airports, hotels, and stadiums for high-density user access.

» **Advancements:**
  - **Wi-Fi 6E Access Points**: Operate on the 6 GHz band, providing faster speeds and lower latency.
  - **Mesh Systems**: Use multiple access points to create a unified, seamless wireless network.

» **Challenges:**
  - **Interference**: Overlapping signals from multiple access points can cause connectivity issues.
  - **Scalability**: Expanding wireless networks without performance degradation can be challenging.

Emerging Trends in Networking Hardware

1. **Software-Defined Networking (SDN)**: SDN separates the control plane from the data plane in network devices, allowing centralized control and automation of networking hardware.
2. **5G-Ready Networking Hardware**: Devices optimized for 5G networks offer ultra-low latency and high-speed connectivity, enabling new applications like autonomous vehicles and smart cities.
3. **Green Networking**: Energy-efficient hardware designs aim to reduce power consumption and environmental impact.
4. **IoT-Specific Devices**: Networking hardware tailored for IoT environments focuses on low power consumption, extended range, and high device density.
5. **Edge Networking Devices**: Hardware that supports edge computing processes data closer to the source, reducing latency and bandwidth usage.

f. Emerging Trends in Hardware

The landscape of hardware technologies is evolving at an unprecedented pace, driven by the demands of a digital-first world. As ICT systems expand to accommodate growing data needs, heightened connectivity, and innovative applications, emerging trends in hardware design and functionality are reshaping how we interact with and utilize technology. These advancements not only enhance performance and efficiency but also address critical challenges related to scalability, security, and environmental sustainability.

1. Internet of Things (IoT) Devices

The Internet of Things (IoT) has become one of the most transformative forces in modern hardware development, embedding sensors, processors, and communication technologies into everyday objects. IoT devices enable these objects to collect, process, and share data autonomously, creating interconnected systems that enhance efficiency, convenience, and decision-making.

- » **Key Features of IoT Devices:**
  - **Sensors**: Capture real-time data such as temperature, motion, humidity, or pressure.
  - **Embedded Processors**: Analyze collected data locally, reducing dependency on external systems.
  - **Connectivity**: Utilize wireless communication protocols like Wi-Fi, Bluetooth, Zigbee, or cellular networks.
- » **Examples of IoT Applications**:
  - **Smart Home Appliances**: Devices like smart thermostats, lighting systems, and security cameras provide automation and remote control.
  - **Wearable Health Trackers**: Devices such as fitness bands and smartwatches monitor vital signs and activity levels.
  - **Industrial Automation**: IoT sensors in manufacturing processes optimize production efficiency and predictive maintenance.
- » **Impact**: IoT devices have revolutionized industries by enabling real-time monitoring, reducing operational costs, and improving user experiences. Their adoption is growing rapidly, with billions of devices connected globally.
- » **Challenges**:
  - **Data Security**: IoT systems are vulnerable to cyberattacks due to their extensive connectivity.
  - **Interoperability**: Ensuring seamless communication between devices from different manufacturers remains a challenge.
  - **Energy Efficiency**: IoT devices often operate in remote locations, necessitating low-power consumption and efficient energy sources.

2. Wearable Technologies

Wearable technologies have emerged as a prominent category of ICT hardware, designed to be worn on the body to deliver real-time information, enhance user interaction, and support specialized applications. These devices integrate compact sensors, processors, and communication modules into ergonomic designs, offering both convenience and functionality.

- » **Types of Wearable Devices:**
  - **Smartwatches and Fitness Trackers**: Monitor health metrics such as heart rate, sleep patterns, and activity levels.
  - **AR Glasses**: Provide augmented reality experiences, overlaying digital information on the real-world view.
  - **Wearable Medical Devices**: Track critical health parameters like blood glucose levels, blood pressure, and ECG in healthcare settings.
  - **Wearable Cameras**: Used in sports, law enforcement, and journalism for hands-free video recording.
- » **Applications**:
  - **Healthcare**: Wearable devices are widely used in telemedicine and remote patient monitoring, empowering healthcare providers with continuous data.
  - **Sports and Fitness**: Athletes use wearables to track performance metrics and optimize training regimens.
  - **Entertainment and Gaming**: AR glasses and haptic wearables create immersive experiences for gamers and media consumers.
- » **Advancements**:
  - **Miniaturization**: Wearable components are becoming smaller, lighter, and more energy-efficient.
  - **AI Integration**: Devices now incorporate machine learning algorithms for predictive analytics and personalized insights.
  - **Flexible Displays**: Emerging designs include bendable screens that conform to the shape of the wearer.
- » **Challenges**:
  - **Battery Life**: Wearables require efficient energy sources to sustain continuous operation.
  - **Privacy Concerns**: The collection of sensitive personal data raises concerns about misuse and data breaches.
  - **Ergonomics**: Designing devices that are comfortable for prolonged use is a critical consideration.

3. Edge Computing Devices

Edge computing represents a paradigm shift in ICT hardware, emphasizing localized data processing near the source of data generation rather than relying solely on centralized cloud systems. Edge devices, such as edge servers and gateways, bring computation and analytics closer to end-users, reducing latency, conserving bandwidth, and enabling real-time decision-making.

» **Key Features**:

- **Localized Processing**: Perform computations at or near the data source, such as sensors or IoT devices.
- **Low Latency**: Minimize delays in data transmission and processing, crucial for time-sensitive applications.
- **Scalability**: Support distributed systems that can grow dynamically as data demands increase.

» **Applications**:

- **Smart Cities**: Edge devices power applications like traffic management, public safety systems, and smart utilities.
- **Healthcare**: Enable real-time analysis of medical data in telehealth environments.
- **Autonomous Vehicles**: Process sensor data locally to support navigation and safety functions.
- **Industrial IoT**: Facilitate predictive maintenance and automation in manufacturing environments.

» **Advantages**:

- **Reduced Bandwidth Usage**: Decreases dependency on transmitting large volumes of data to centralized servers.
- **Enhanced Security**: Limits the exposure of sensitive data by keeping it closer to the source.
- **Improved Reliability**: Edge systems can operate independently, ensuring continued functionality during network disruptions.

» **Challenges**:

- **Integration**: Ensuring compatibility between edge devices and centralized systems.
- **Energy Requirements**: Balancing high performance with energy efficiency in localized devices.
- **Cost**: Deploying edge infrastructure involves significant initial investment.

4. Quantum Computing Hardware

Quantum computing is an emerging technology poised to revolutionize ICT by solving complex problems far beyond the capabilities of classical computers. Quantum hardware operates using quantum bits (qubits) that exploit the principles of quantum mechanics, such as superposition and entanglement, to perform computations.

» **Features of Quantum Hardware**:

- **Qubits**: Unlike classical bits, which are either 0 or 1, qubits can represent both states simultaneously, enabling exponential increases in computational power.

- **Quantum Gates**: Perform operations on qubits to process information.
- **Cryogenic Systems**: Quantum processors often require extremely low temperatures to maintain qubit coherence.

» **Potential Applications**:

- **Cryptography**: Revolutionizing encryption and decryption methods.
- **Drug Discovery**: Simulating molecular interactions to accelerate pharmaceutical research.
- **Optimization Problems**: Solving logistical challenges in transportation, supply chain management, and finance.
- **Artificial Intelligence**: Enhancing machine learning algorithms with faster and more efficient computations.

» **Challenges**:

- **Scalability**: Building quantum systems with a large number of stable qubits.
- **Error Correction**: Managing decoherence and noise that affect quantum calculations.
- **Accessibility**: Current quantum systems are expensive and primarily accessible through research institutions or cloud-based platforms.

5. Energy-Efficient Hardware

The growing emphasis on environmental sustainability has driven the development of energy-efficient hardware solutions that minimize power consumption and reduce electronic waste. These advancements align with global efforts to combat climate change and promote green ICT practices.

» **Key Features**:

- **Low-Power Processors**: CPUs and GPUs optimized for minimal energy use without sacrificing performance.
- **Renewable Materials**: Use of recyclable and biodegradable components in hardware design.
- **Dynamic Power Management**: Systems that adjust energy usage based on workload demands.

» **Applications**:

- **Data Centers**: Transitioning to energy-efficient servers and cooling systems to reduce carbon footprints.
- **Consumer Electronics**: Incorporating energy-saving modes in devices like laptops, smartphones, and home appliances.
- **IoT Systems**: Designing low-power sensors and devices for long-term deployments in remote areas.

- **Advancements**:
  - **Solar-Powered Devices**: Hardware that harnesses solar energy for operation in off-grid environments.
  - **AI for Energy Optimization**: Machine learning algorithms that predict and manage energy usage in real-time.
  - **Carbon-Neutral Manufacturing**: Adopting sustainable practices in hardware production to minimize environmental impact.
- **Challenges**:
  - **Balancing Cost and Efficiency**: Developing affordable energy-efficient hardware remains a significant hurdle.
  - **Lifecycle Management**: Ensuring the recyclability of components and reducing e-waste.

g. Challenges in Hardware Development

The evolution of hardware technology has driven remarkable advancements in Information and Communication Technology (ICT) systems. However, alongside these innovations, the development, deployment, and maintenance of hardware components present significant challenges that require continuous attention and innovation. These challenges stem from economic, technical, environmental, and security-related factors, affecting the accessibility, sustainability, and reliability of ICT infrastructure.

1. Cost

The development and production of advanced hardware often involve substantial financial investments in research, manufacturing, and quality assurance. This high cost creates barriers to accessibility, particularly in economically disadvantaged regions or organizations with limited budgets.

Factors Contributing to High Costs

- **Research and Development (R&D)**: Cutting-edge hardware requires significant investment in design, prototyping, and testing.
- **Manufacturing Complexity**: Advanced components, such as microprocessors or GPUs, require sophisticated manufacturing processes involving high-precision machinery and cleanroom environments.
- **Specialized Materials**: Many modern hardware components, like semiconductor chips, rely on rare or expensive materials.
- **Supply Chain Dependencies**: Disruptions in the global supply chain can increase the cost of raw materials and components, as seen during events like the COVID-19 pandemic.

Impact

- Limited adoption of advanced technologies in low-income regions.
- Slower technology penetration in education, healthcare, and small businesses, where cost-sensitive decisions are paramount.

Potential Solutions

- **Government Subsidies and Grants**: Programs to support the adoption of hardware in underserved areas.
- **Open Hardware Initiatives**: Promoting collaborative, cost-effective development of hardware designs.
- **Advancements in Manufacturing**: Scaling up production processes to reduce per-unit costs.

2. Scalability

Overview

As ICT systems expand in complexity and scale, ensuring that hardware components can grow seamlessly with system demands presents a significant challenge. Scalability issues often arise in environments like data centers, smart cities, and industrial automation systems, where hardware must support increasing workloads and connectivity requirements.

Challenges in Scalability

- **Hardware Compatibility**: Integrating new hardware with existing systems often requires substantial adjustments, leading to inefficiencies and higher costs.
- **Infrastructure Limitations**: Power supplies, cooling systems, and network bandwidth may not scale efficiently with growing hardware demands.
- **Resource Allocation**: Balancing computational resources across hardware components to prevent bottlenecks.

Impact

- Reduced system performance and efficiency in large-scale deployments.
- Increased operational costs due to frequent upgrades and replacements.

Potential Solutions

- **Modular Hardware Designs**: Components that can be easily upgraded or expanded as system requirements grow.
- **Edge Computing**: Offloading processing tasks to localized edge devices reduces the burden on centralized hardware systems.
- **Dynamic Resource Allocation**: Leveraging software-defined hardware to optimize scalability and resource distribution.

3. Obsolescence

Rapid technological advancements often render existing hardware obsolete within a short period, creating challenges for users, manufacturers, and the environment. The lifecycle of hardware components has become increasingly compressed, especially in consumer electronics and high-tech industries.

Factors Driving Obsolescence

- » **Technological Innovations**: Newer, more efficient hardware designs outpace older models in terms of performance and energy efficiency.
- » **Software Requirements**: Modern software often demands higher hardware specifications, forcing users to upgrade.
- » **Consumer Trends**: Market-driven demand for the latest devices accelerates product turnover.

Impact

- » Increased electronic waste (e-waste) due to discarded hardware.
- » Financial burdens on users and organizations needing frequent upgrades.
- » Inefficiencies in industries dependent on long-term hardware usage.

Potential Solutions

- » **Sustainable Design Practices**: Creating hardware with longer lifespans and upgradeable components.
- » **Refurbishing and Recycling**: Establishing programs to refurbish outdated hardware and recycle valuable materials.
- » **Support for Legacy Systems**: Providing extended software support for older hardware to extend its usability.

4. Security

Hardware vulnerabilities are a critical concern in ICT systems, as they can serve as entry points for cyberattacks. Security flaws in hardware can compromise entire systems, leading to data breaches, operational disruptions, and financial losses.

Common Hardware Security Issues

- » **IoT Vulnerabilities**: Many IoT devices have limited security features, making them attractive targets for hackers.
- » **Firmware Exploits**: Malicious actors can exploit weaknesses in hardware firmware to gain unauthorized access.
- » **Side-Channel Attacks**: Advanced techniques that exploit physical characteristics of hardware, such as power consumption or electromagnetic emissions, to extract sensitive information.

Impact

- » Compromise of critical systems, including healthcare, finance, and infrastructure.
- » Loss of consumer trust in technology providers.
- » Increased costs associated with mitigating security breaches and implementing protective measures.

Potential Solutions

» **Hardware-Based Security Features**: Incorporating elements like Trusted Platform Modules (TPMs) and secure boot processes into hardware designs.

» **Regular Firmware Updates**: Ensuring that devices receive timely updates to patch vulnerabilities.

» **IoT Security Standards**: Establishing industry-wide standards for securing IoT devices and networks.

5. Environmental Impact

The environmental consequences of hardware development, including resource extraction, energy consumption, and waste generation, pose significant challenges for sustainability.

Environmental Challenges

» **Resource Depletion**: Mining for materials like rare earth metals and lithium has significant ecological impacts.

» **Energy Usage**: Hardware manufacturing and operation contribute to greenhouse gas emissions.

» **E-Waste**: Discarded hardware often ends up in landfills, releasing toxic substances into the environment.

Impact

» Long-term environmental degradation and loss of biodiversity.

» Increased costs for managing and disposing of e-waste.

Potential Solutions

» **Eco-Friendly Materials**: Using renewable and biodegradable materials in hardware design.

» **Energy-Efficient Manufacturing**: Adopting processes that minimize energy consumption and emissions.

» **Circular Economy Models**: Encouraging the reuse, refurbishment, and recycling of hardware components.

6. Supply Chain Disruptions

Global supply chains for hardware components are highly complex and vulnerable to disruptions from geopolitical tensions, natural disasters, and pandemics. These disruptions can delay production, increase costs, and create shortages.

Challenges in Supply Chains

» **Dependency on Specific Regions**: Many critical components, such as semiconductors, are manufactured in a few countries, creating bottlenecks.

- » **Logistics Challenges**: Delays in shipping and transportation can hinder timely delivery of components.
- » **Trade Restrictions**: Tariffs and embargoes can limit access to essential hardware materials.

Impact

- » Delayed product launches and missed market opportunities.
- » Increased costs for manufacturers and end-users.

Potential Solutions

- » **Regional Diversification**: Expanding manufacturing capabilities to multiple regions to reduce dependency.
- » **Inventory Management**: Maintaining strategic reserves of critical components to mitigate shortages.
- » **Technological Alternatives**: Developing new materials and manufacturing techniques to reduce reliance on scarce resources.

2. SOFTWARE

Software is a vital component of Information and Communication Technology (ICT) systems, encompassing the programs and instructions that enable hardware to perform specific tasks. It acts as an intermediary between users and machines, translating human commands into a language that computers can understand and execute. Without software, even the most advanced hardware would be inert, unable to deliver functionality or value. Software has evolved to become one of the most dynamic and versatile components of ICT, influencing nearly every facet of modern life. From operating systems that govern hardware functionality to sophisticated applications that solve complex problems, software drives innovation and efficiency across industries.

Categories of Software

1. System Software

System software is a fundamental component of any Information and Communication Technology (ICT) environment, acting as the backbone that governs the operation of hardware and provides essential services for other types of software. It establishes an interface between users and machines, enabling seamless interaction and coordination. Without system software, hardware would lack the ability to execute tasks, and application software would have no foundation upon which to operate. System software is designed to manage and optimize the hardware resources of a computing device, ensuring efficiency, reliability, and security. It encompasses a wide array of programs that facilitate hardware functionality, coordinate processes, and deliver a user-friendly environment.

Components of System Software

1. **Operating Systems (OS)**: The operating system is the core component of system software, responsible for managing hardware resources, providing user interfaces, and enabling the execution of applications. Operating systems act as the central hub that coordinates all activities within a computer system.

» **Key Functions of Operating Systems:**

- **Process Management:** Handles the scheduling and execution of multiple processes, ensuring optimal performance and preventing conflicts.
- **Memory Management:** Allocates and deallocates memory to processes, ensuring efficient utilization of available resources.
- **File System Management:** Organizes and secures data stored on disk drives, enabling easy access and retrieval.
- **User Interface:** Provides graphical (GUI) or command-line interfaces (CLI) that allow users to interact with the system.

» **Examples of Operating Systems:**

- **Windows:** Known for its widespread adoption in personal and professional environments, Windows offers a user-friendly interface and extensive compatibility with software and hardware.
- **macOS:** Developed by Apple, macOS is renowned for its seamless integration with Apple devices and focus on design aesthetics and security.
- **Linux:** A versatile, open-source operating system widely used in servers, development environments, and embedded systems due to its customizability and robust performance.

2. **Utility Programs:** Utility programs are specialized tools within system software that perform specific tasks to maintain, optimize, or protect the performance of a computing device. These tools enhance the functionality of the operating system by addressing targeted needs.

» **Types of Utility Programs:**

- **Antivirus Software:** Detects and eliminates malware, protecting the system from security threats.
- **Disk Cleanup Tools:** Remove unnecessary files and optimize storage space.
- **Backup Utilities:** Enable data backup and recovery to safeguard against data loss.
- **System Monitoring Tools:** Provide real-time insights into resource usage, such as CPU and memory utilization.

» **Role of Utility Programs:**

- Enhance system performance through optimization and regular maintenance.
- Provide diagnostic tools for identifying and resolving issues.
- Improve security and data integrity.

Functions of System Software

System software performs a variety of functions that are essential for the operation of ICT systems. These functions ensure that hardware resources are utilized efficiently and that users can interact with their devices effectively.

1. **Resource Management**
   - Allocates CPU, memory, and storage resources to running processes.
   - Manages resource contention and prioritizes tasks to ensure smooth system performance.
   - Monitors hardware usage to prevent overloading or bottlenecks.
2. **File System Management**
   - Organizes data into directories and files, making it easier to store, retrieve, and manage information.
   - Provides access control mechanisms to safeguard data from unauthorized access.
   - Supports different file systems, such as NTFS, FAT32, and ext4, to cater to diverse requirements.
3. **Device Control**
   - Interfaces with peripheral devices such as printers, scanners, monitors, and input devices.
   - Provides drivers that enable hardware components to communicate with the operating system.
   - Ensures compatibility between hardware and software components.
4. **System Security**
   - Implements authentication mechanisms, such as passwords and biometric verification.
   - Protects the system from unauthorized access and malicious software.
   - Encrypts data to prevent eavesdropping and data breaches.
5. **Error Detection and Handling**
   - Monitors system operations to detect errors or failures.
   - Provides tools for diagnosing and resolving hardware or software issues.
   - Ensures system stability by isolating faulty processes or components.
6. **User Interface Management**
   - Offers graphical or command-line interfaces for user interaction.
   - Provides accessibility features to accommodate users with disabilities.
   - Simplifies navigation and system control through intuitive design.

Impact of System Software

System software plays a pivotal role in the functionality and reliability of ICT systems. Its impact extends across various domains, influencing user experience, operational efficiency, and technological innovation.

1. **Ensuring Reliability**: System software provides a stable environment for hardware and

applications to operate. By managing resources and processes effectively, it minimizes crashes, freezes, and other disruptions, ensuring consistent performance.

2. **Enabling Scalability**: Modern operating systems and utility programs are designed to support scalability, accommodating growing workloads and expanding hardware infrastructures. This capability is crucial for enterprise environments and cloud-based systems.
3. **Enhancing User Experience**: Through user-friendly interfaces and accessibility features, system software simplifies interaction with complex ICT systems. Customizable settings and preferences allow users to tailor their experiences to their needs.
4. **Driving Innovation**: Advances in system software drive technological progress by enabling new applications and capabilities. For example:
    - Virtualization technologies allow multiple operating systems to run on a single machine.
    - Real-time operating systems (RTOS) power critical systems in industries such as healthcare and aerospace.
5. **Supporting Integration**: System software facilitates integration with diverse hardware and software ecosystems, enabling interoperability and collaboration. This capability is essential in environments with heterogeneous systems, such as IoT networks and multi-cloud platforms.

2. Application Software

Application software refers to programs and tools designed to perform specific tasks or solve particular problems for end-users. Unlike system software, which manages hardware and system-level operations, application software directly addresses the needs of individuals or organizations, enhancing productivity, enabling communication, and delivering entertainment. Application software is the most visible aspect of ICT systems, offering user-centric solutions across various domains. With advancements in technology, the scope and capabilities of application software have expanded dramatically, addressing increasingly complex tasks in professional, personal, and industrial contexts. From word processing and video conferencing to specialized software for healthcare and design, application software has become an indispensable part of modern life.

Categories of Application Software

1. Productivity Tools

Productivity tools are among the most widely used types of application software, designed to help users efficiently complete everyday tasks such as document creation, data analysis, and project management.

» **Examples**:

- **Microsoft Office Suite**: Includes Word (word processing), Excel (spreadsheets), and PowerPoint (presentation creation).
- **Google Workspace**: Offers cloud-based alternatives like Docs, Sheets, and Slides, enabling real-time collaboration.

» **Features**:

- **Templates**: Pre-designed formats for documents, spreadsheets, and presentations that save time.
- **Collaboration Tools**: Enable multiple users to work on the same document simultaneously.
- **Automation**: Functions like formulas in spreadsheets and macros in word processors streamline repetitive tasks.

» **Applications**: Productivity tools are essential in workplaces, educational institutions, and personal projects, enabling users to manage their tasks effectively and meet deadlines.

2. Communication Tools

Communication tools facilitate interaction and collaboration among individuals and teams, regardless of physical location. These tools have become increasingly important in the age of remote work and global connectivity.

» **Examples**:

- **Zoom**: Popular for video conferencing, webinars, and virtual meetings.
- **Slack**: A team collaboration platform that combines messaging, file sharing, and project tracking.

» **Features**:

- **Real-Time Communication**: Includes video, voice, and instant messaging options.
- **File Sharing**: Allows users to share documents, images, and videos seamlessly.
- **Integrations**: Many communication tools integrate with other software like calendars, project management systems, and cloud storage platforms.

» **Applications**:

- **Business**: Remote teams use these tools to coordinate projects and communicate efficiently.
- **Education**: Facilitates virtual classrooms and e-learning environments.
- **Social Interaction**: Connects friends and family across distances.

3. Entertainment Software

Entertainment software is designed for leisure and enjoyment, encompassing games, media players, and streaming platforms. These applications cater to diverse audiences, offering immersive experiences and relaxation.

» **Examples**:

- **Gaming Software**: Titles like Fortnite, Minecraft, and The Sims provide interactive gaming experiences.
- **Streaming Platforms**: Services like Netflix, Spotify, and YouTube allow users to consume video and audio content.

» **Features:**
- **Customization:** Games often allow users to personalize characters, settings, and gameplay.
- **Content Recommendations:** Streaming platforms use algorithms to suggest content based on user preferences.
- **Multiplayer Options:** Many games offer online multiplayer modes for social interaction.

» **Applications:** Entertainment software is widely used for relaxation, creative expression, and connecting with others through shared experiences.

4. Specialized Industry Software

Specialized industry software is tailored to meet the unique needs of specific professional fields or industries. These applications often address complex workflows, regulatory requirements, and industry-specific challenges.

» **Examples:**
- **Healthcare:**
  - ◊ **Electronic Health Record (EHR) Systems:** Manage patient data, appointments, and billing processes.
  - ◊ Examples: Epic, Cerner.
- **Design and Engineering:**
  - ◊ **Computer-Aided Design (CAD) Software:** Tools like AutoCAD and SolidWorks are used for creating detailed engineering designs and prototypes.
- **Finance:**
  - ◊ **Accounting Software:** Applications like QuickBooks and Xero simplify financial management and reporting.

» **Features:**
- **Customizable Workflows:** Designed to align with specific industry practices.
- **Data Security:** Compliance with industry standards for protecting sensitive information.
- **Advanced Analytics:** Provide insights and trends relevant to the field.

» **Applications:** Specialized software is crucial in fields like healthcare, engineering, and finance, where precision, compliance, and efficiency are paramount.

Functions of Application Software

Application software performs several critical functions that enhance user experience and efficiency:

1. **Task-Specific Solutions:**
   - Each application is tailored to address a specific need, whether it is writing a report, analyzing data, or managing a patient's medical records.

2. **User Interfaces:**
   - Application software provides interfaces designed for usability and accessibility, often including graphical elements, customization options, and multilingual support.
3. **Integration with Other Software:**
   - Many applications are designed to work seamlessly with other software, enabling users to share data and automate workflows across platforms.
4. **Enhanced Functionality:**
   - Applications often include advanced features like artificial intelligence, machine learning, and cloud integration to expand their capabilities.

Impact of Application Software

The impact of application software extends across personal, professional, and industrial domains, driving progress and innovation in countless areas:

1. **Empowering Productivity:**
   - Productivity tools enable users to achieve their goals efficiently, whether it's completing a school project or running a business operation.
2. **Facilitating Global Communication:**
   - Communication applications break down geographical barriers, enabling collaboration and connectivity on a global scale.
3. **Driving Innovation**:
   - Specialized software fosters creativity and innovation in industries like healthcare, design, and entertainment, pushing the boundaries of what's possible.
4. **Enhancing Quality of Life**:
   - Entertainment software provides leisure and relaxation, contributing to mental well-being.
   - Applications like fitness trackers and meditation apps promote healthier lifestyles.
5. **Economic Growth**:
   - The application software industry generates billions of dollars in revenue annually, creating jobs and driving technological advancements.

3. Middleware

Middleware plays a crucial role in modern ICT systems, acting as an intermediary layer that facilitates communication, integration, and data management between different software applications. In complex environments, where multiple applications or services must work together seamlessly, middleware ensures compatibility and streamlines operations. By bridging the gap between diverse systems, middleware enhances functionality, improves efficiency, and enables scalability. Middleware is particularly vital in enterprise settings, IoT ecosystems, and distributed computing environments, where the integration of disparate systems is essential for smooth operations.

Components of Middleware

Middleware comprises various components, each designed to address specific integration and communication challenges in ICT systems. These components ensure that data flows smoothly and consistently across platforms, regardless of differences in architecture or protocols.

1. Database Middleware

Database middleware connects applications to databases, enabling efficient data access, manipulation, and management. It simplifies interactions between the application layer and the underlying database, ensuring seamless operations.

- » **Key Features:**
  - **Query Translation:** Converts application-level queries into database-compatible commands.
  - **Connection Pooling:** Manages a pool of database connections to optimize performance and resource usage.
  - **Transaction Management:** Ensures data integrity during multi-step processes by implementing mechanisms like commit and rollback.
- » **Applications:**
  - Data-driven applications such as e-commerce platforms and financial systems.
  - Real-time analytics tools that require rapid data access.

2. Message-Oriented Middleware (MOM)

Message-oriented middleware facilitates asynchronous communication between distributed systems, ensuring that messages are delivered reliably even in environments with network latency or interruptions.

- » **Key Features:**
  - **Message Queues:** Store messages temporarily until the receiving system is ready to process them.
  - **Publish-Subscribe Model:** Allows multiple systems to subscribe to specific message types, promoting efficient data dissemination.
  - **Fault Tolerance:** Ensures message delivery in case of system failures or network disruptions.
- » **Applications:**
  - Event-driven architectures, such as IoT ecosystems where sensors and devices exchange data.
  - Microservices-based applications requiring seamless communication between components.

3. Application Servers

Overview

Application servers provide an environment for running specific applications, often in enterprise settings. They act as a middleware layer that manages application logic and integrates it with backend systems and databases.

» **Key Features:**
  - **Load Balancing:** Distributes workloads across servers to ensure consistent performance.
  - **Security Services:** Implements authentication, authorization, and encryption for secure data exchange.
  - **Scalability:** Supports dynamic scaling to accommodate fluctuating user demands.

» **Applications:**
  - Web applications, such as e-commerce sites and content management systems.
  - Enterprise solutions like ERP (Enterprise Resource Planning) and CRM (Customer Relationship Management) platforms.

Functions of Middleware

Middleware performs several critical functions that enable the integration and smooth operation of complex ICT systems. These functions enhance the flexibility, reliability, and scalability of interconnected applications.

1. Integration

Middleware ensures compatibility between applications with varying architectures, programming languages, and communication protocols. It acts as a translator, allowing disparate systems to work together seamlessly.

» **Example:**
  - In an e-commerce platform, middleware integrates the payment gateway, inventory management system, and shipping service.

2. Data Exchange

Middleware facilitates the smooth transfer of data across platforms, ensuring consistency and accuracy. It manages the formatting, routing, and delivery of data, even in distributed environments.

» **Example:**
  - Middleware in a healthcare system enables data exchange between electronic health record (EHR) systems and diagnostic tools, ensuring accurate and timely information.

3. Transaction Management

Middleware ensures the consistency and reliability of multi-step processes, particularly in scenarios involving financial transactions or complex workflows.

» **Features:**
  - **Atomicity:** Ensures that all steps in a transaction are completed successfully or rolled back entirely.

- **Consistency**: Maintains data integrity across systems.
- **Durability**: Guarantees that completed transactions are preserved even in the event of system failures.

» **Example**:

- In banking applications, middleware manages fund transfers between accounts, ensuring that all steps are completed accurately.

Applications of Middleware

Middleware is widely used across various industries and ICT systems to streamline operations and enhance efficiency. Key applications include:

1. **Enterprise Resource Planning (ERP) Systems**: Middleware integrates the various modules of an ERP system, such as finance, human resources, and supply chain management, ensuring smooth data flow and consistency.
2. **Customer Relationship Management (CRM) Platforms**: Middleware connects CRM tools with other enterprise systems like marketing automation software and sales databases, enabling comprehensive customer insights.
3. **IoT Ecosystems**: Middleware facilitates communication between IoT devices, sensors, and cloud platforms, enabling real-time data processing and decision-making.
4. **Cloud Computing**: Middleware supports the integration of cloud-based applications with on-premise systems, ensuring hybrid cloud functionality.
5. **Big Data Analytics**: Middleware enables data aggregation and processing from multiple sources, supporting real-time analytics and business intelligence.

Impact of Middleware

Middleware significantly impacts the functionality, scalability, and adaptability of ICT systems, making it a cornerstone of modern technology infrastructure.

1. **Simplifies Complex Operations**: By acting as a central hub for integration, middleware reduces the complexity of managing interconnected systems. This simplification enables organizations to focus on innovation rather than operational challenges.
2. **Enables Scalability**: Middleware provides the flexibility to add or modify system components without disrupting existing workflows. This capability is critical in environments with growing user bases or expanding operations.
3. **Enhances Interoperability**: Middleware bridges the gap between legacy systems and modern applications, ensuring that organizations can leverage their existing investments while adopting new technologies.
4. **Supports Real-Time Processing**: Middleware enables real-time data processing and decision-making, which is essential in applications like IoT, financial transactions, and emergency response systems.

5. **Drives Digital Transformation**: Middleware facilitates the seamless integration of emerging technologies, such as AI, machine learning, and blockchain, into existing ICT ecosystems, accelerating digital transformation initiatives.

Advancements in Software Development

The field of software development has undergone rapid advancements in recent years, driven by emerging technologies and evolving methodologies. These innovations are transforming how software is designed, delivered, and utilized, enabling new capabilities, and redefining the role of software in various industries. This section explores four key advancements in software development: Artificial Intelligence (AI) and Machine Learning (ML), Cloud Computing, DevOps and Agile Development, and Open-Source Software. Each represents a significant leap forward, enhancing efficiency, scalability, and collaboration while addressing contemporary challenges.

1. Artificial Intelligence (AI) and Machine Learning (ML)

Artificial Intelligence (AI) and Machine Learning (ML) are revolutionizing software development by enabling systems to learn, adapt, and make intelligent decisions based on data. These technologies mimic human cognitive processes, allowing software to perform tasks that traditionally require human intervention. From automating mundane tasks to solving complex problems, AI and ML are at the forefront of innovation.

» **Applications**:

1. **Natural Language Processing (NLP)**: Powers virtual assistants like Siri, Alexa, and Google Assistant, enabling them to understand and respond to human language. NLP is also used in chatbots, sentiment analysis, and language translation.
2. **Predictive Analytics**: Employed in marketing to predict consumer behavior, in finance to forecast stock trends, and in healthcare to anticipate disease outbreaks and treatment outcomes.
3. **Autonomous Systems**: Facilitates the development of self-driving cars, drones, and robotic process automation (RPA) for tasks such as data entry and invoice processing.

» **Impact**:

- **Efficiency**: Automates repetitive and time-consuming tasks, freeing up human resources for more strategic roles.
- **Personalization**: Delivers customized user experiences, such as tailored recommendations in e-commerce or personalized learning paths in education.
- **Enhanced Decision-Making**: Provides data-driven insights that improve accuracy and reduce risks in business, healthcare, and public policy.

» **Challenges**:

- Data privacy concerns due to the extensive data collection required for training models.
- Bias in algorithms, which can lead to unfair outcomes in applications like hiring or credit scoring.

- The need for specialized expertise and computational resources.

2. Cloud Computing

Cloud computing has fundamentally changed the way software is developed, deployed, and accessed. By shifting from on-premise installations to web-based services, cloud computing offers unprecedented flexibility and scalability. It allows users to access software and resources from anywhere with an internet connection, reducing dependence on local hardware.

» **Benefits**:

1. **Scalability**: Resources can be scaled up or down dynamically based on user demand, ensuring cost efficiency and optimal performance.
2. **Accessibility**: Enables users to access applications and data from any device, fostering mobility and remote collaboration.
3. **Cost Efficiency**: Reduces the need for expensive hardware and IT infrastructure, as resources are hosted and maintained by cloud providers.

» **Examples**:

- **Software as a Service (SaaS)**: Platforms like Salesforce, Microsoft 365, and Google Workspace deliver software applications over the internet.
- **Infrastructure as a Service (IaaS)**: Providers such as Amazon Web Services (AWS), Microsoft Azure, and Google Cloud Platform offer virtualized computing resources on demand.

» **Impact**:

- **Global Collaboration**: Facilitates teamwork across geographic locations through shared platforms and tools.
- **Data Sharing**: Promotes seamless data exchange and integration across systems and applications.
- **Innovation**: Accelerates the development and deployment of new software solutions by providing developers with on-demand resources and tools.

» **Challenges**:

- Security risks, including data breaches and unauthorized access.
- Downtime or reduced performance during service outages.
- Regulatory compliance issues, particularly concerning data storage and processing across international borders.

3. DevOps and Agile Development

Modern software development practices like DevOps and Agile Development emphasize collaboration, flexibility, and continuous delivery. These methodologies are designed to address the challenges of traditional development models, such as slow timelines and communication gaps between development and operations teams.

» **DevOps**:
  - Combines development and IT operations to create a unified workflow.
  - Encourages automation of repetitive tasks, such as code integration, testing, and deployment.
  - Promotes a culture of collaboration, breaking down silos between teams.

» **Agile Development**:
  - Focuses on iterative processes, where development is broken into small, manageable cycles (sprints).
  - Allows teams to adapt to changing requirements and feedback quickly.
  - Prioritizes delivering a minimum viable product (MVP) early, with continuous improvements over time.

» **Impact**:
  - **Accelerated Time-to-Market**: Enables faster delivery of software updates and new features.
  - **Improved Software Quality**: Continuous testing and feedback loops help identify and resolve issues early in the development process.
  - **Enhanced Collaboration**: Encourages cross-functional teamwork, fostering innovation and alignment with business goals.

» **Challenges**:
  - Requires a cultural shift within organizations, which can be difficult to achieve.
  - Demands skilled personnel familiar with DevOps and Agile methodologies.
  - Complexities in automating certain tasks and integrating diverse tools.

4. Open-Source Software

The open-source movement has democratized software development, making high-quality tools and frameworks accessible to everyone. Open-source software is developed collaboratively by communities of contributors who share their work freely. This approach has led to rapid innovation, increased transparency, and cost savings for organizations.

» **Benefits**:
  1. **Cost Savings**: Organizations can use and customize open-source solutions without incurring licensing fees.
  2. **Transparency**: Open access to source code allows developers to identify and address vulnerabilities, enhancing security.
  3. **Rapid Innovation**: Collaborative contributions from a global community drive faster improvements and feature additions.

» **Examples**:

- **Linux**: A widely used open-source operating system known for its stability and scalability.
- **Apache Hadoop**: A framework for distributed storage and processing of big data, critical for industries like finance, healthcare, and technology.

» **Impact**:

- **Accessibility**: Levels the playing field by providing tools to small businesses, startups, and developing regions.
- **Community Engagement**: Fosters a culture of collaboration and shared knowledge.
- **Customization**: Enables organizations to tailor solutions to their specific needs.

» **Challenges**:

- Lack of formal support compared to proprietary software.
- Potential compatibility issues with other systems or software.
- Dependency on community contributors for updates and bug fixes.

Challenges in Software Development

Software development is a cornerstone of technological progress, driving innovation and enabling solutions across industries. However, the field faces significant challenges that developers, organizations, and policymakers must address to ensure the reliability, security, and usability of software systems. These challenges stem from the complexities of evolving technologies, increasing cyber threats, and the growing need for seamless integration across diverse platforms.

1. Security Vulnerabilities

As software becomes more interconnected and data-intensive, protecting against cyberattacks and data breaches has become a critical priority. Security vulnerabilities in software can lead to unauthorized access, data theft, and disruption of services, posing risks to individuals, businesses, and governments.

» **Key Challenges**:

- **Complex Attack Vectors**: Cybercriminals exploit vulnerabilities such as weak authentication, poor encryption, and insecure APIs.
- **Evolving Threat Landscape**: The rise of sophisticated threats like ransomware, phishing, and zero-day attacks demands proactive defense mechanisms.
- **Legacy Systems**: Older software often lacks modern security features, making it a target for attackers.

» **Impact**:

- Financial losses due to fraud, fines, and downtime.
- Reputational damage to organizations affected by breaches.
- Compromised user trust and confidence in digital systems.

» **Potential Solutions:**

- **Adopting Secure Development Practices:** Incorporate security from the outset using methodologies like DevSecOps.
- **Regular Updates and Patches:** Address vulnerabilities promptly to minimize exposure.
- **User Education:** Raise awareness about safe practices, such as avoiding phishing scams and using strong passwords.

2. Interoperability

In today's interconnected world, software systems often need to communicate with diverse platforms and devices. Ensuring seamless integration and compatibility between systems with different architectures, protocols, and standards is a persistent challenge.

» **Key Challenges:**

- **Diverse Technology Stacks:** Variability in programming languages, frameworks, and APIs complicates integration.
- **Lack of Standardization:** Inconsistent standards across industries and regions hinder interoperability.
- **Data Format Incompatibilities:** Differences in data structures and formats can lead to errors and inefficiencies.

» **Impact:**

- Increased development time and costs due to custom integrations.
- Limited scalability and flexibility in systems that cannot communicate effectively.
- Reduced user satisfaction when software fails to integrate with existing tools.

» **Potential Solutions:**

- **Open Standards and Protocols:** Encourage the adoption of universally accepted standards, such as JSON for data exchange or RESTful APIs for communication.
- **Middleware Solutions:** Use middleware to bridge gaps between disparate systems and facilitate data flow.
- **Modular Design:** Develop software with a focus on modularity and reusability to simplify integration.

3. Rapid Evolution

The fast-paced evolution of technology creates pressure on software developers to keep up with emerging trends, tools, and user expectations. Staying ahead of the curve while maintaining software quality and stability is a significant challenge.

» **Key Challenges:**

- **Technological Obsolescence:** Frequent updates to programming languages, frameworks, and platforms can render existing software outdated.

- **User Expectations**: Modern users demand intuitive interfaces, real-time updates, and high-performance applications.
- **Time-to-Market Pressure:** Organizations often prioritize speed over thorough testing, increasing the risk of bugs and vulnerabilities.

» **Impact**:

- High development and maintenance costs due to frequent updates.
- Reduced product lifespan as older technologies are phased out.
- Developer burnout from the constant need to learn and adapt.

» **Potential Solutions**:

- **Agile Development Practices**: Use iterative methodologies to adapt quickly to changes without sacrificing quality.
- **Continuous Learning**: Invest in training and upskilling for development teams to stay current with new technologies.
- **Future-Proofing**: Design software with scalability and flexibility in mind to accommodate future advancements.

Emerging Challenges

In addition to the core challenges mentioned above, software development faces emerging issues as technology continues to evolve:

1. **Artificial Intelligence and Machine Learning**:
   - Developing ethical AI systems that avoid bias and ensure transparency.
   - Ensuring the interpretability and reliability of AI-driven decision-making processes.
2. **Sustainability**:
   - Reducing the environmental impact of software systems by optimizing energy consumption in data centers and cloud environments.
3. **Regulatory Compliance**:
   - Navigating complex and evolving regulations such as GDPR (General Data Protection Regulation) and HIPAA (Health Insurance Portability and Accountability Act) to ensure software meets legal requirements.

3. Networks

Networks are the foundation of modern Information and Communication Technology (ICT), enabling interconnected devices and systems to communicate, share resources, and collaborate efficiently. They form the infrastructure that supports the exchange of data across various domains, from small office environments to global-scale applications like the Internet. By connecting devices through communication protocols, networks facilitate real-time interactions, information access, and resource sharing, making them indispensable in today's digital era.

Types of Networks

Networks can be categorized based on their geographical scope, communication technology, and purpose. Below are the key types of networks that form the backbone of ICT systems:

1. **Local Area Networks (LANs)**: A Local Area Network (LAN) connects devices within a confined area, such as an office, school, or home. LANs are widely used for resource sharing, including printers, files, and internet connections.
   » **Features**:
      - Typically spans a single building or campus.
      - Offers high-speed data transfer (up to several gigabits per second).
      - Requires minimal infrastructure, often using Ethernet cables or Wi-Fi for connectivity.

   » **Applications**:
      - Office environments for connecting computers, printers, and servers.
      - Schools and universities for sharing resources among students and staff.
      - Homes for linking smart devices and personal computers.
2. **Wide Area Networks (WANs)**: A Wide Area Network (WAN) spans a larger geographical area, often connecting multiple LANs. WANs are essential for enabling communication and resource sharing across cities, countries, or even continents.
   » **Features**:
      - Uses public or private communication links, such as leased telephone lines, satellite connections, or fiber-optic cables.
      - Often relies on technologies like Multiprotocol Label Switching (MPLS) and Virtual Private Networks (VPNs) for secure communication.

   » **Applications**:
      - Connecting branch offices of a multinational corporation.
      - Supporting government communication networks across regions.
      - Providing internet access through Internet Service Providers (ISPs).
3. **Wireless Networks**: Wireless networks eliminate the need for physical cables by using radio waves, infrared, or microwaves to connect devices. They provide mobility and flexibility, making them ideal for modern, dynamic environments.
   » **Types of Wireless Networks**:
      - **Wi-Fi**: Used in homes, offices, and public spaces for high-speed internet access.
      - **Bluetooth**: Facilitates short-range communication between devices, such as smartphones and wireless headphones.
      - **5G Networks**: The latest generation of mobile communication, offering ultra-fast speeds, low latency, and massive device connectivity.

» **Applications**:
   - Smart cities for connecting IoT devices and infrastructure.
   - Public transportation systems for real-time updates and Wi-Fi access.
   - Industrial automation for wireless control of machinery and sensors.

4. **The Internet**: The internet is the global network that connects millions of private and public networks, enabling access to information and services worldwide. It is the most extensive example of a WAN and has transformed how individuals and organizations communicate and operate.

   » **Features**:
   - Supports a wide range of applications, including web browsing, email, e-commerce, and social media.
   - Uses standardized protocols like TCP/IP for data exchange.
   - Accessible through various technologies, such as fiber optics, DSL, and satellite.

   » **Impact**:
   - Revolutionized communication through instant messaging, video calls, and online collaboration tools.
   - Enabled the proliferation of e-learning, telemedicine, and e-governance services.
   - Facilitated the growth of the digital economy, including online retail and digital marketing.

Network Protocols

Protocols are standardized rules that govern data exchange between devices on a network. They ensure reliable, secure, and efficient communication, regardless of the devices or systems involved.

1. **TCP/IP (Transmission Control Protocol/Internet Protocol)**: The backbone of the internet and most networks, TCP/IP enables reliable communication by breaking data into packets and reassembling them at the destination.
2. **HTTP/HTTPS (Hypertext Transfer Protocol/Secure)**: Used for transferring web pages and content over the internet. HTTPS adds encryption for secure communication.
3. **FTP (File Transfer Protocol)**: Facilitates the transfer of files between computers on a network.
4. **SMTP (Simple Mail Transfer Protocol)**: Enables email transmission across networks.
5. **SNMP (Simple Network Management Protocol)**: Used for monitoring and managing network devices, such as routers and switches.

Cloud Networks

Cloud technology has revolutionized networking by offering scalable, on-demand access to computing resources and services over the Internet. Cloud networks eliminate the need for traditional infrastructure, providing flexibility and cost-efficiency.

» **Key Features**:
  - **Scalability**: Resources can be dynamically allocated to meet changing demands.
  - **Accessibility**: Data and applications can be accessed from anywhere, enabling remote work and collaboration.
  - **Integration**: Cloud networks seamlessly integrate with existing IT infrastructure, supporting hybrid and multi-cloud environments.

» **Examples**:
  - **Content Delivery Networks (CDNs)**: Accelerate the delivery of web content by caching data at geographically distributed servers.
  - **Virtual Private Clouds (VPCs)**: Offer isolated environments within a public cloud for enhanced security and control.

» **Applications**:
  - Supporting global businesses with distributed teams and operations.
  - Hosting scalable applications, such as e-commerce platforms and streaming services.
  - Enabling real-time analytics and decision-making in industries like finance and healthcare.

Integration of Secure, High-Speed Networks

Modern ICT systems rely on secure, high-speed networks to enable real-time communication, collaboration, and data exchange. Key considerations include:

1. **Security**:
   - Implementing encryption protocols to protect data in transit.
   - Using firewalls, intrusion detection systems, and VPNs to safeguard network integrity.
   - Regularly updating network devices and protocols to address vulnerabilities.
2. **Speed**:
   - Adopting high-speed technologies like 5G, fiber optics, and low-latency satellite networks.
   - Optimizing network performance through load balancing and traffic prioritization.
3. **Global Connectivity**:
   - Leveraging cloud networks and the internet to connect people and systems worldwide.
   - Supporting cross-border collaboration in education, healthcare, and business.

Impact of Networks on ICT

Networks have a transformative impact on ICT systems, enabling innovation and progress across various domains:

» **Real-Time Communication**: Facilitates instant messaging, video conferencing, and live data sharing, driving productivity and collaboration.

- » **Data-Driven Decision-Making**: Supports the collection and analysis of data from distributed sources, enhancing insights and outcomes.
- » **Scalability and Flexibility**: Allows businesses to scale operations without significant infrastructure investment, adapting to changing demands.
- » **Innovation in Industries**: Powers advancements in IoT, cloud computing, and artificial intelligence, transforming sectors like healthcare, agriculture, and logistics.

4. Data

Data serves as the cornerstone of any Information and Communication Technology (ICT) system, representing raw facts and figures that, when processed, generate meaningful insights and drive decision-making. In the digital age, the value of data has grown exponentially, influencing every aspect of business, governance, healthcare, and education. Efficient data management, processing, storage, and security are critical to leveraging the full potential of ICT systems.

Key Aspects of Data in ICT

1. Data Types

Data can be classified into various types based on its structure and format. These classifications determine how data is processed, stored, and analyzed.

1. **Structured Data**: Structured data is highly organized and formatted into predefined structures such as tables, rows, and columns. It is easily searchable and can be stored in relational databases.
   - » **Examples**:
     - Customer data in a CRM system, such as names, addresses, and purchase histories.
     - Financial records in spreadsheets, including revenues, expenses, and budgets.
   - » **Applications**:
     - Transaction processing in banking systems.
     - Inventory management in supply chains.
2. **Unstructured Data**: Unstructured data lacks a predefined format, making it more complex to process and analyze. It includes a wide range of formats, such as text documents, images, audio, and video.
   - » **Examples**:
     - Social media posts, including tweets and Facebook updates.
     - Multimedia files, such as photographs and recorded lectures.
   - » **Applications**:
     - Sentiment analysis in marketing.
     - Content indexing in multimedia libraries.

3. **Big Data**: Big Data refers to large and complex datasets generated from various sources, often characterized by high volume, velocity, and variety. Advanced tools and technologies are required to store, process, and analyze Big Data.
   - » **Examples**:
     - Sensor data from IoT devices.
     - Web logs and clickstream data from e-commerce platforms.
   - » **Applications**:
     - Predictive analytics in healthcare and finance.
     - Fraud detection in banking.

2. Data Processing

Data processing involves transforming raw data into meaningful information. This process includes several stages, each critical to deriving value from data.

1. **Input**: Data is collected from various sources, such as sensors, user interactions, or external databases. Effective input mechanisms ensure the accuracy and completeness of collected data.
   - » **Sources**:
     - IoT devices and sensors in smart cities.
     - User-entered data in web forms.
     - API integrations with external systems.
2. **Processing**: During this stage, data is manipulated using algorithms, software tools, and statistical methods. Processing may include filtering, aggregation, transformation, and analysis.
   - » **Techniques**:
     - Data cleaning to remove errors or inconsistencies.
     - Statistical analysis to identify trends and patterns.
     - Machine learning algorithms for predictive modeling.
3. **Output**: Processed data is presented in a user-friendly format, such as visualizations, reports, or dashboards. This step ensures that the insights are actionable and accessible to decision-makers.
   - » **Formats**:
     - Graphs and charts in business intelligence dashboards.
     - Summary reports for management reviews.

3. Data Storage and Retrieval

Efficient storage and retrieval systems are essential for managing data in ICT systems. These systems ensure that data is readily available, organized, and secure.

1. **Relational Databases**: Relational databases use structured query language (SQL) to organize data into tables with predefined relationships.

- **Examples:**
  - MySQL, PostgreSQL, and Oracle Database.
- **Applications:**
  - Customer relationship management (CRM) systems.
  - Enterprise resource planning (ERP) solutions.

2. **Non-Relational Databases:** Non-relational or NoSQL databases handle unstructured or semi-structured data, offering flexibility and scalability.
   - **Examples:**
     - MongoDB, Cassandra, and Couchbase.
   - **Applications:**
     - Content management systems for websites.
     - Real-time analytics in e-commerce.
3. **Data Warehouses:** Data warehouses are centralized repositories designed for storing and analyzing large datasets. They support advanced analytics and business intelligence applications.
   - **Examples:**
     - Snowflake, Amazon Redshift, and Google BigQuery.
   - **Applications:**
     - Generating sales forecasts and market trends.
     - Consolidating data from multiple sources for unified reporting.

4. Data Security

Ensuring the confidentiality, integrity, and availability of data is a fundamental aspect of ICT systems. Data security involves protecting data from unauthorized access, corruption, or loss.

1. **Key Security Measures:**
   - **Encryption:** Converts data into an unreadable format to protect it during transmission and storage.
   - **Firewalls:** Act as barriers to prevent unauthorized access to networks.
   - **Intrusion Detection Systems (IDS):** Monitor network activity to detect and respond to potential threats.
2. **Applications:**
   - Securing financial transactions in online banking.
   - Protecting patient records in healthcare systems.
   - Safeguarding intellectual property in research and development.
3. **Emerging Challenges:**
   - **Cyberattacks:** Increasing sophistication of threats such as ransomware and phishing.

- **Data Privacy**: Compliance with regulations like GDPR and CCPA to protect user data.
- **Cloud Security**: Ensuring the security of data stored and processed in cloud environments.

Impact of Data in ICT

1. **Decision-Making**: Data-driven decision-making has transformed industries by providing insights that enhance efficiency, reduce costs, and identify opportunities. For example, predictive analytics helps businesses anticipate customer needs and optimize inventory management.
2. **Innovation**: Access to vast amounts of data fuels innovation in fields like artificial intelligence, machine learning, and IoT. These technologies rely on data to develop smarter systems and applications.
3. **Efficiency**: Effective data management streamlines operations, improves resource allocation, and reduces waste. In manufacturing, for instance, real-time data from sensors enables predictive maintenance, minimizing downtime.
4. **Global Connectivity**: Data facilitates global communication and collaboration by enabling platforms like video conferencing, file sharing, and e-learning. This connectivity has become especially crucial in remote work and education scenarios.

5. People

The human element is an integral component of any Information and Communication Technology (ICT) system, as technology is fundamentally designed to serve and empower people. From end-users who rely on ICT tools for daily tasks to professionals who develop and maintain these systems, and decision-makers who shape policies and strategies, the human role in ICT cannot be overstated. The effectiveness and impact of ICT systems depend largely on the interaction, expertise, and vision of the people involved.

Key Roles in ICT

1. End-Users

End-users are individuals or groups who interact directly with ICT systems and tools to perform tasks, access services, or achieve specific objectives. They are the ultimate beneficiaries of ICT and play a pivotal role in shaping how technology evolves through their needs, feedback, and usage patterns.

- **Types of End-Users**:
  1. **Students and Educators**:
     - Use e-learning platforms, digital libraries, and collaborative tools to facilitate education and skill development.
     - Examples: Platforms like Coursera, Zoom, and Google Classroom have transformed traditional learning environments into digital ecosystems.
  2. **Professionals:**
     - Rely on ICT tools for productivity, communication, and collaboration.

- Examples: Cloud-based tools like Microsoft Teams, Slack, and project management software like Trello enable remote work and team coordination.

3. **General Public:**

- Access ICT for various purposes, such as entertainment (streaming platforms), online shopping, and social networking.

» **Impact**:

- End-users drive demand for innovation, encouraging developers to create user-centric technologies.
- Their feedback helps improve system usability, security, and accessibility.

2. ICT Professionals

ICT professionals are responsible for designing, developing, maintaining, and optimizing ICT systems. They possess the technical expertise required to manage the complex infrastructure that powers modern technology.

» **Key Roles and Responsibilities**:

1. **System Administrators**:
   - **Role**: Manage and maintain ICT infrastructure, ensuring its reliability, performance, and security.
   - **Responsibilities**:
     - ◊ Monitor system performance and troubleshoot issues.
     - ◊ Implement security measures, such as firewalls and access controls.
     - ◊ Schedule regular backups and disaster recovery plans.
   - **Impact**: Ensure the seamless operation of ICT systems, minimizing downtime and disruptions.
2. **Software Developers**:
   - **Role**: Design, code, and maintain software applications that meet user requirements.
   - **Responsibilities**:
     - ◊ Write and debug code to create functional and user-friendly applications.
     - ◊ Collaborate with stakeholders to understand system requirements.
     - ◊ Update and improve software to address bugs or incorporate new features.
   - **Impact**: Enable the creation of tools and platforms that drive productivity, innovation, and connectivity.
3. **Data Analysts:**
   - **Role**: Extract insights from data to support decision-making and strategy formulation.

- **Responsibilities**:
  - ◊ Analyze structured and unstructured data using tools like Excel, Python, or Tableau.
  - ◊ Identify trends, patterns, and opportunities for improvement.
  - ◊ Present findings through visualizations and reports.
- **Impact**: Drive data-driven decisions, enhancing organizational efficiency and competitiveness.

4. **Network Engineers:**
   - **Role**: Design, implement, and maintain communication networks that facilitate data exchange.
   - **Responsibilities**:
     - ◊ Set up and configure routers, switches, and other network hardware.
     - ◊ Monitor network performance and optimize for speed and reliability.
     - ◊ Implement security protocols to protect against cyber threats.
   - **Impact**: Ensure the availability and reliability of communication networks, enabling real-time connectivity and collaboration.

3. Decision-Makers and Policymakers

Decision-makers and policymakers play a strategic role in guiding the adoption and implementation of ICT systems. Their vision, strategies, and policies determine how ICT is leveraged for social and economic development.

» **Roles and Contributions:**

1. **Organizational Leaders**:
   - ◊ Develop ICT strategies aligned with business goals.
   - ◊ Invest in technology to enhance operational efficiency and innovation.
   - ◊ Ensure compliance with industry standards and regulations.
2. **Government Policymakers**:
   - ◊ Formulate national ICT policies to promote digital inclusion, infrastructure development, and economic growth.
   - ◊ Implement e-governance initiatives to improve public service delivery.
   - ◊ Address challenges such as the digital divide and cybersecurity.

» **Impact:**

- Leaders shape the trajectory of technological advancement, influencing sectors like healthcare, education, and commerce.
- Their policies ensure equitable access to technology, fostering social and economic empowerment.

The Interconnected Roles of People in ICT

The roles of end-users, professionals, and decision-makers are interconnected, creating a feedback loop that drives ICT innovation and effectiveness:

1. **End-Users Influence Development**:
   » Feedback from end-users informs developers about system usability and desired features.
   » Usage patterns guide the prioritization of updates and enhancements.
2. **ICT Professionals Enable Usability**:
   » Developers, administrators, and analysts create and maintain systems that cater to user needs.
   » Their expertise ensures that systems are secure, reliable, and scalable.
3. **Decision-Makers Guide Adoption**:
   » Policies and strategies set by leaders determine the focus areas for ICT deployment.
   » Investment decisions influence the accessibility and sustainability of ICT infrastructure.

Challenges Faced by People in ICT

Despite their critical roles, individuals in the ICT ecosystem face several challenges:

1. **Digital Skills Gap**:
   » Many end-users lack the technical skills required to maximize the benefits of ICT tools.
   » ICT professionals face the constant need to upskill to keep pace with technological advancements.
2. **Cybersecurity Threats**:
   » Users and professionals alike are vulnerable to cyberattacks, which can compromise data and system integrity.
   » Decision-makers must balance the need for accessibility with robust security measures.
3. **Equitable Access**:
   » Policymakers must address disparities in access to ICT resources, particularly in underserved or rural areas.

The Human-Centered Future of ICT

As ICT systems evolve, the role of people will remain central to their success. Emerging trends such as artificial intelligence, automation, and IoT will transform how individuals interact with technology, creating opportunities for enhanced productivity and innovation.

1. **User-Centric Design:**
   » Developers will focus on creating intuitive interfaces and personalized experiences.
   » Accessibility features will ensure inclusivity for users with diverse abilities.
2. **Empowered Professionals**:

- » Advancements in tools and platforms will enable ICT professionals to work more efficiently and creatively.
- » Continuous learning opportunities will help them adapt to changing technologies.

3. **Informed Policymaking**:
   - » Leaders will use data analytics and predictive modeling to make evidence-based decisions.
   - » Policies will emphasize sustainability, equity, and resilience in ICT systems.

Integration of Components

The integration of ICT components—hardware, software, networks, data, and people—forms the backbone of modern digital ecosystems. This interconnectedness is what enables ICT to power a wide range of applications across industries, fostering innovation, improving efficiency, and enhancing communication. By working in harmony, these components create systems that are greater than the sum of their parts, delivering transformative outcomes in areas like education, healthcare, business, and governance.

Integration of ICT Components in Applications

The integration of hardware, software, networks, data, and people ensures the seamless functionality of ICT systems, driving innovation and improving outcomes. Below are key examples illustrating this integration in real-world applications:

1. Education

In education, ICT has transformed traditional teaching and learning processes by enabling e-learning platforms and digital classrooms. These systems integrate various ICT components to create a cohesive learning environment.

- » **Hardware**:
  - Devices such as laptops, tablets, and interactive whiteboards serve as tools for students and educators.
- » **Software**:
  - Learning Management Systems (LMS) like Moodle and Google Classroom provide platforms for course delivery, assessments, and collaboration.
- » **Networks**:
  - Internet connectivity ensures access to online resources, virtual classrooms, and cloud-based tools.
- » **Data**:
  - Course materials, student performance metrics, and attendance records are collected, processed, and analyzed to enhance learning outcomes.
- » **People**:
  - Teachers, students, and administrators interact with the system, creating a collaborative

educational experience.

» **Impact**:

- Promotes accessibility through distance learning.
- Enables personalized education through data analytics.
- Encourages global collaboration among educators and learners.

2. Healthcare

The healthcare sector has embraced ICT to deliver innovative solutions that improve patient care, streamline operations, and enhance diagnostic accuracy.

» **Hardware**:

- Devices such as telemedicine kiosks, wearable health monitors, and imaging equipment capture and relay patient data.

» **Software**:

- Applications like Electronic Health Records (EHRs) and diagnostic tools facilitate data management and clinical decision-making.

» **Networks**:

- High-speed networks enable real-time communication between patients and healthcare providers, supporting telemedicine and remote monitoring.

» **Data**:

- Patient histories, diagnostic results, and treatment plans are stored and analyzed for evidence-based care.

» **People**:

- Healthcare professionals, patients, and administrative staff interact with ICT systems to improve efficiency and outcomes.

» **Impact**:

- Expands access to healthcare services, especially in remote areas.
- Enhances diagnostic accuracy through AI-powered tools.
- Reduces costs by optimizing resource allocation and patient management.

3. Business and Industry

ICT is a driving force in modern businesses, enabling automation, enhancing communication, and supporting data-driven decision-making.

» **Hardware**:

- Servers, IoT devices, and smart manufacturing tools optimize operations.

» **Software**:

- Enterprise Resource Planning (ERP) systems, Customer Relationship Management (CRM) platforms, and analytics tools streamline workflows.

» **Networks**:

- Cloud-based networks facilitate remote work and global collaboration.

» **Data**:

- Customer insights, supply chain metrics, and financial records guide strategy and operations.

» **People**:

- Employees, managers, and customers engage with ICT systems to improve efficiency and satisfaction.

» **Impact**:

- Enables remote work and flexible business models.
- Enhances decision-making through real-time data analytics.
- Drives innovation in product design and service delivery.

Emerging Trends in ICT Components

The field of ICT is continuously evolving, with advancements redefining the roles and capabilities of its components. Emerging trends highlight the need for adaptability and innovation to leverage these changes effectively.

1. Edge Computing

Overview

Edge computing brings data processing closer to its source, reducing latency and bandwidth usage. This trend addresses the growing need for real-time analytics and decision-making in applications like IoT and autonomous vehicles.

» **Applications**:

- Smart cities: Traffic management systems process data locally for real-time updates.
- Healthcare: Wearable devices analyze patient data at the edge, enabling immediate responses.

» **Impact**:

- Improves speed and efficiency by minimizing reliance on centralized data centers.
- Enhances privacy and security by keeping sensitive data closer to its origin.

2. 5G Networks

Overview

5G networks represent the next generation of mobile communication, offering unprecedented speeds, low latency, and massive device connectivity. This technology is a cornerstone for enabling

IoT ecosystems, smart devices, and high-bandwidth applications.

» **Applications:**
- Autonomous vehicles: Facilitates real-time communication between vehicles and infrastructure.
- Augmented and Virtual Reality (AR/VR): Enables seamless streaming and interaction for immersive experiences.

» **Impact:**
- Supports large-scale IoT deployments.
- Enhances the quality and reliability of mobile and remote services.

3. Artificial Intelligence (AI) and Machine Learning (ML)

Overview

AI and ML are revolutionizing software capabilities by enabling predictive analysis, automation, and personalization. These technologies enhance the functionality of ICT systems, making them smarter and more adaptive.

» **Applications:**
- E-commerce: AI-driven recommendations improve customer experience.
- Healthcare: ML algorithms analyze medical images for early diagnosis.

» **Impact:**
- Reduces operational costs through automation.
- Enhances decision-making with predictive insights.

4. Blockchain

Overview

Blockchain technology is redefining data security and transparency in ICT systems by creating tamper-proof, decentralized ledgers. This innovation is particularly impactful in sectors that require trust and accountability.

» **Applications:**
- Finance: Enables secure, transparent cryptocurrency transactions.
- Supply Chain: Tracks goods throughout their lifecycle to ensure authenticity.

» **Impact:**
- Enhances trust in digital transactions.
- Streamlines processes by eliminating intermediaries.

B. The Role of ICT in Community Education

Information and Communication Technology (ICT) plays a transformative role in community education, reshaping how knowledge is disseminated, accessed, and utilized. By leveraging ICT, communities can overcome traditional barriers to education, such as geographical constraints, limited resources, and socio-economic disparities. ICT has enabled inclusive, flexible, and scalable educational models that address the diverse learning needs of individuals and groups across the globe.

1. Enabling Access to Education

Access to quality education has long been a challenge for underserved communities due to constraints like inadequate infrastructure, teacher shortages, and financial limitations. Information and Communication Technology (ICT) has emerged as a transformative force, breaking down these barriers and democratizing education. By leveraging ICT, learners across the globe can access a wealth of knowledge and resources, irrespective of their location, socio-economic status, or physical abilities.

Distance Learning

Distance learning is one of the most significant contributions of ICT to education, enabling learners to access high-quality content from anywhere in the world. This mode of learning is particularly beneficial for students in remote or underserved areas who lack access to traditional educational facilities.

» **Key Features of Distance Learning**:
  - Provides flexibility for learners to study at their own pace and convenience.
  - Reduces the need for physical infrastructure, making education more cost-effective.
  - Facilitates global learning, connecting students with educators and institutions worldwide.

» **Applications**:
  - **Online Education Platforms**: Platforms like Coursera, edX, and Khan Academy offer courses across disciplines, catering to learners of all ages and backgrounds. These platforms provide video lectures, quizzes, and assignments, ensuring a comprehensive learning experience.
  - **Virtual Classrooms**: Tools like Zoom and Microsoft Teams support real-time interaction between educators and students, replicating the traditional classroom experience in a virtual setting.

» **Impact**:
  - Expands access to education for marginalized groups, including women, rural populations, and refugees.
  - Empowers individuals to acquire new skills and qualifications, enhancing employability and career growth.
  - Promotes lifelong learning by offering flexible learning opportunities for working professionals and adult learners.

E-Learning Platforms

E-learning platforms leverage ICT to create structured digital learning environments where educators can deliver content, track progress, and engage with students. These platforms cater to diverse educational needs, from formal schooling to skill development and corporate training.

» **Key Components of E-Learning Platforms**:
- **Learning Management Systems (LMS)**: Platforms like Moodle, Google Classroom, and Blackboard provide educators with tools to design courses, upload materials, and assess student performance. LMS platforms also enable interactive features like discussion boards and peer reviews.
- **Mobile Learning Apps**: Apps like BYJU's and Duolingo make education accessible to learners with limited access to computers. BYJU's focuses on K-12 education, offering engaging video lessons, while Duolingo specializes in language learning through gamification.

» **Advantages**:
- Personalization: E-learning platforms use analytics and algorithms to tailor content to individual learning preferences and progress.
- Scalability: Digital platforms can accommodate a large number of users simultaneously, making them ideal for massive open online courses (MOOCs).
- Cost-Effectiveness: Reduces expenses related to textbooks, transportation, and physical infrastructure.

» **Impact**:
- Revolutionizes education delivery by integrating multimedia elements, interactive tools, and real-time feedback.
- Bridges the gap between learners and educators, especially during crises like the COVID-19 pandemic.
- Encourages self-paced learning, fostering independence and self-motivation among students.

Community Resource Centers

Community resource centers equipped with ICT infrastructure play a pivotal role in extending educational opportunities to underserved populations. These centers act as hubs where individuals can access technology, digital libraries, and training programs.

» **Features of ICT-Equipped Centers**:
- Provide free or subsidized access to computers, the internet, and software tools.
- Offer training programs on digital literacy, basic computing skills, and vocational courses.
- Facilitate group learning and collaborative activities through shared resources.

» **Examples**:

- **Digital Villages:** Initiatives like India's Common Service Centers (CSCs) provide ICT services to rural communities, including e-learning and skill development programs.
- **Public Libraries with ICT Facilities**: Many libraries now integrate digital resources, enabling users to access e-books, online journals, and educational databases.

» **Impact**:

- Bridges the digital divide by providing equitable access to technology and learning resources.
- Empowers communities with knowledge and skills, fostering socio-economic development.
- Supports lifelong learning by offering resources for diverse age groups and educational levels.

Inclusive Education

ICT has significantly contributed to making education more inclusive by addressing the needs of learners with disabilities and other challenges. Specialized tools and resources ensure that all individuals, regardless of their abilities, can access and benefit from education.

» **Key ICT Tools for Inclusive Education**:

- **Assistive Technologies:**
  ◊ Screen readers like JAWS and NVDA help visually impaired learners access digital content.
  ◊ Speech-to-text applications such as Dragon NaturallySpeaking convert spoken words into written text, aiding students with motor impairments or dyslexia.
- **Accessible Learning Platforms:**
  ◊ Many e-learning platforms now incorporate accessibility features, such as closed captions, adjustable text sizes, and high-contrast modes.

» **Benefits**:

- Promotes equal opportunities by ensuring that education is accessible to all.
- Reduces barriers to learning for individuals with disabilities or special needs.
- Encourages inclusive classroom environments where diverse learners can collaborate and thrive.

» **Impact**:

- Enhances confidence and independence among learners with disabilities.
- Expands access to specialized knowledge and resources, empowering marginalized groups.
- Drives innovation in educational technology, inspiring the development of more inclusive tools and platforms.

2. Enhancing Teaching and Learning

ICT has redefined the teaching and learning process, introducing tools, methodologies, and resources that make education more interactive, engaging, and effective. By integrating technology into educational settings, educators can enrich the learning experience, cater to diverse learning styles, and foster collaboration and innovation among students.

Interactive Learning Tools

Interactive learning tools leverage ICT to create immersive and engaging educational experiences, moving beyond traditional teaching methods to stimulate curiosity and foster deeper understanding.

1. Virtual Reality (VR) and Augmented Reality (AR)

» **Applications**:

- **Virtual Field Trips**: VR allows students to visit historical sites, explore the solar system, or dive into the ocean—all without leaving the classroom.
- **Simulated Experiments**: AR enables students to interact with 3D models of complex systems, such as the human body or chemical reactions, enhancing comprehension in science and engineering.

» **Benefits**:

- Provides hands-on experience in a safe and controlled environment.
- Encourages active participation and engagement by immersing learners in the subject matter.
- Bridges the gap between theory and practice, making abstract concepts more tangible.

» **Examples**:

- Google Expeditions offers virtual reality tours of historical landmarks, ecosystems, and cultural sites.
- Apps like Human Anatomy VR allow medical students to explore the human body in three dimensions.

2. Interactive Whiteboards and Multimedia Presentations

» **Applications**:

- **Interactive Whiteboards**: Allow teachers to write, draw, and interact with digital content, such as videos, charts, and animations, during lessons.
- **Multimedia Presentations**: Combine text, images, videos, and audio to present information in an engaging and multi-sensory format.

» **Benefits**:

- Accommodates various learning styles (visual, auditory, and kinesthetic) to improve comprehension and retention.
- Encourages student participation through quizzes, polls, and collaborative exercises.

- Provides flexibility for educators to adapt lessons in real-time based on student feedback.

» **Examples**:

- SMART Boards enable real-time interaction with digital content, enhancing classroom engagement.
- Tools like Prezi and Canva help educators create visually appealing and dynamic presentations.

Personalized Learning

Personalized learning uses ICT to tailor the educational experience to each learner's unique needs, preferences, and pace. This approach empowers students to take ownership of their learning journey while enabling educators to address individual strengths and weaknesses.

1. AI-Powered Platforms

» **Applications**:

- AI analyzes learner behavior and performance, identifying areas where additional support or resources are needed.
- Platforms recommend personalized study materials, practice tests, and learning strategies based on individual progress.

» **Benefits**:

- Provides immediate feedback, enabling students to address gaps in understanding.
- Encourages self-paced learning, accommodating diverse levels of readiness and ability.
- Increases engagement by delivering content that aligns with students' interests and goals.

» **Examples**:

- Khan Academy uses AI to suggest personalized practice exercises and tutorials.
- EdTech platforms like DreamBox Learning adapt math lessons to suit individual learning trajectories.

2. Adaptive Learning Systems

» **Applications**:

- Adjust difficulty levels and pace based on real-time performance data.
- Create dynamic learning pathways that evolve as students achieve mastery in specific topics.

» **Benefits**:

- Reduces frustration by ensuring that students are neither overwhelmed nor under-challenged.
- Enhances learning efficiency by focusing on areas that require improvement.

- Supports learners with varying abilities, including those with special needs.

» **Examples:**

- Platforms like ALEKS (Assessment and Learning in Knowledge Spaces) use adaptive learning algorithms to personalize math and science education.
- Duolingo adjusts language lessons based on user progress and proficiency.

Collaborative Learning

Collaborative learning leverages ICT to foster teamwork, knowledge sharing, and problem-solving among students. By creating virtual spaces for interaction, ICT enables learners to collaborate on projects and exchange ideas, both locally and globally.

1. Group Projects and Discussions

» **Applications:**

- Platforms like Microsoft Teams and Zoom enable students to work together on assignments, share resources, and discuss ideas in real time.
- Features like breakout rooms and shared documents facilitate seamless collaboration.

» **Benefits:**

- Encourages communication and teamwork, essential skills for the modern workforce.
- Provides opportunities for peer learning, where students can benefit from each other's strengths and perspectives.
- Reduces geographical and cultural barriers, enabling global collaboration.

» **Examples:**

- Students use Google Docs to co-author essays and presentations in real time.
- Zoom breakout rooms allow small group discussions during virtual classes.

3. Empowering Educators

Educators play a crucial role in shaping the future by imparting knowledge, skills, and values to learners. Information and Communication Technology (ICT) has revolutionized the educational landscape, empowering educators with tools and resources to enhance teaching effectiveness and improve instructional quality. By integrating ICT into their workflows, educators can deliver engaging lessons, streamline administrative tasks, and access opportunities for professional growth.

Professional Development

Professional development is essential for educators to stay updated with the latest teaching methodologies, technological advancements, and subject matter expertise. ICT provides numerous avenues for continuous learning, enabling teachers to refine their skills and adapt to evolving educational needs.

1. Online Courses and Webinars

- **Overview:** Online platforms offer educators the flexibility to learn at their own pace, exploring a wide range of topics from pedagogy to technology integration.
- **Examples:**
  - **LinkedIn Learning:** Offers courses on classroom management, instructional design, and leadership skills.
  - **Coursera and edX:** Provide access to university-level courses in education and technology, often leading to certifications.
  - **TeacherTube:** A video-sharing platform specifically designed for educators, featuring instructional videos, teaching tips, and professional development resources.
- **Benefits:**
  - Enhances educators' knowledge of innovative teaching techniques, such as flipped classrooms and project-based learning.
  - Provides exposure to global best practices in education.
  - Fosters a sense of community among educators through discussion forums and collaborative projects.

2. Platforms for Continuous Learning

- **Applications:**
  - ICT platforms offer microlearning modules, podcasts, and webinars to address specific teaching challenges.
  - Conferences and virtual summits provide opportunities to interact with experts and peers.
- **Examples:**
  - **TeachThought:** A resource hub offering articles, courses, and podcasts on modern teaching strategies.
  - **Education Week Webinars:** Focused on topics like educational equity, technology integration, and curriculum design.
- **Impact:**
  - Encourages lifelong learning among educators, promoting adaptability and resilience.
  - Bridges knowledge gaps, ensuring that teachers are well-prepared to meet diverse learner needs.

Digital Content Creation

ICT empowers educators to create, adapt, and share digital content, fostering creativity and collaboration in lesson planning and delivery. These tools enable teachers to design interactive and engaging educational materials tailored to their students' needs.

1. Tools for Content Creation

» **Applications**:
- **Canva**: Helps educators design visually appealing infographics, posters, and presentations.
- **Google Slides**: Facilitates collaborative creation of multimedia-rich presentations.
- **Kahoot**: A game-based learning platform that allows teachers to create interactive quizzes and polls.

» **Benefits**:
- Simplifies the process of creating high-quality educational resources.
- Encourages creativity by allowing customization of content to suit different learning styles.
- Enhances student engagement through the use of multimedia elements.

2. Open Educational Resources (OER)

OER are freely available teaching materials that educators can access, modify, and share. These resources include lesson plans, worksheets, textbooks, and multimedia content.

» **Examples**:
- **OER Commons**: A digital library offering resources across various subjects and grade levels.
- **Khan Academy**: Provides instructional videos and exercises for free use by educators worldwide.

» **Benefits**:
- Reduces the cost of instructional materials, making education more accessible.
- Encourages knowledge sharing and collaboration among educators.
- Promotes adaptability, as resources can be tailored to local contexts and diverse learning needs.

» **Impact**:
- Empowers teachers to deliver customized and inclusive lessons.
- Supports innovation in curriculum design by providing access to diverse perspectives and ideas.

Efficient Classroom Management

ICT tools streamline administrative tasks, enabling educators to focus more on teaching and student engagement. By automating routine processes, these tools save time and reduce the burden of classroom management.

1. Attendance Tracking and Grading

» **Applications:**

- Tools like **ClassDojo** and **Edmodo** allow educators to monitor attendance, manage assignments, and maintain gradebooks.
- Learning management systems (LMS) like Moodle and Blackboard automate grading for quizzes and assignments.

» **Benefits:**

- Reduces paperwork and administrative workload.
- Provides real-time insights into student performance and attendance patterns.
- Ensures accurate record-keeping, facilitating better communication with parents and administrators.

» **Examples:**

- **PowerSchool:** A student information system that integrates attendance, grading, and reporting features.
- **Google Classroom:** Simplifies assignment submission, feedback, and grading processes.

2. Parent-Teacher Communication

» **Applications:**

- Communication platforms enable seamless interaction between teachers and parents, fostering transparency and collaboration.
- Features include messaging systems, progress reports, and event notifications.

» **Examples:**

- **Remind:** Allows teachers to send updates and reminders to parents and students.
- **Seesaw:** A digital portfolio tool where students can share their work, and parents can monitor progress.

» **Benefits:**

- Builds stronger relationships between educators and families.
- Encourages parental involvement in student learning.
- Promotes a supportive learning environment through regular updates and feedback.

Impact of ICT on Educators

The integration of ICT into teaching empowers educators in several ways:

1. **Enhanced Teaching Effectiveness:**
   - Access to diverse resources and tools enables teachers to deliver engaging and impactful lessons.

- Data analytics provides insights into student performance, allowing for targeted interventions.

2. **Increased Efficiency**:
   - Automation of administrative tasks frees up time for planning and instruction.
   - Digital tools streamline communication and collaboration with students, parents, and colleagues.
3. **Professional Growth**:
   - Opportunities for continuous learning help educators stay at the forefront of pedagogical innovation.
   - Exposure to global teaching practices broadens perspectives and fosters creativity.

Challenges in ICT Adoption for Educators

Despite its benefits, ICT adoption among educators faces several challenges:

1. **Digital Literacy:**
   - Not all educators are equipped with the technical skills required to effectively use ICT tools.
   - Training programs are essential to bridge this gap and build confidence.
2. **Infrastructure Gaps:**
   - Lack of access to reliable internet, devices, and software hinders ICT adoption, particularly in underserved areas.
3. **Time Constraints:**
   - Teachers often struggle to balance ICT integration with existing workloads and curriculum requirements.
4. **Cost of Tools and Resources**:
   - While some ICT tools are free, others require significant investment, limiting access for resource-constrained schools.

4. Bridging the Digital Divide

The digital divide refers to the gap between individuals, households, businesses, and geographic areas at different socio-economic levels concerning their access to, use of, or knowledge of information and communication technologies (ICT). Bridging this divide is essential for fostering inclusive growth, reducing poverty, and enhancing the overall socio-economic development of communities.

Community Outreach Programs

Community outreach programs are instrumental in extending the benefits of ICT to remote and disadvantaged areas. Governments and non-governmental organizations (NGOs) collaborate to establish ICT training centers, enhancing digital literacy and providing access to technology.

» **Establishment of ICT Training Centers**:

- **Purpose**: These centers aim to equip individuals with essential digital skills, enabling them to participate effectively in the digital economy. Training includes basic computer operations, internet navigation, online communication, and the use of various software applications.
- **Implementation**: Centers are strategically located in accessible areas within communities, such as schools, libraries, or community centers. They are staffed by trained professionals who provide instruction and support.
- **Services Offered**:
  - ◊ **Digital Literacy Courses**: Covering topics from basic computer use to advanced IT skills.
  - ◊ **Access to Technology**: Providing computers, tablets, and internet connectivity for public use.
  - ◊ **Workshops and Seminars**: Focusing on topics like cybersecurity, online entrepreneurship, and e-governance services.
- **Impact**:
  - ◊ **Empowerment**: Individuals gain confidence and skills to access information, services, and opportunities online.
  - ◊ **Economic Development**: Enhances employability and opens avenues for income generation through digital means.

» **Government Initiatives**:

» **India's Digital India Campaign**:

- ◊ **Overview**: A flagship program initiated by the Government of India to transform the country into a digitally empowered society and knowledge economy.
- ◊ **Objectives**:
  - **Digital Infrastructure as a Core Utility**: Ensuring high-speed internet availability for all citizens.
  - **Governance and Services on Demand**: Seamless integration of services across departments or jurisdictions.
  - **Digital Empowerment of Citizens**: Universal digital literacy and accessible digital resources.
- ◊ **Key Projects**:
  - **Common Service Centers (CSCs)**: Provide access points for the delivery of essential public utility services, social welfare schemes, and financial services.
  - **BharatNet**: Aims to connect all 250,000 Gram Panchayats in the country with high-speed broadband.

◊ **Other Examples**:

- **Kenya's Pasha Centers**: Digital villages offering ICT services, training, and access to government services.
- **Brazil's Telecentros**: Public spaces providing free access to computers and the internet, along with training in digital skills.

» **NGO Contributions**:

- **One Laptop per Child (OLPC)**: Provides low-cost, durable laptops to children in developing countries to facilitate self-empowered learning.
- **World Computer Exchange**: Focuses on reducing the digital divide by providing refurbished computers and educational resources to youth in developing countries.
- **Local Initiatives**: NGOs often tailor programs to address specific community needs, such as digital literacy for women, youth empowerment through ICT, and supporting small businesses.

Affordable Technology Solutions

Affordability is a significant barrier preventing marginalized populations from accessing ICT. Initiatives that provide low-cost devices and subsidized internet services are crucial in making technology accessible to all.

» **Low-Cost Devices**:

- **Raspberry Pi**:
  - **Description**: An affordable, compact computer designed to promote the teaching of basic computer science in schools and developing countries.
  - **Applications**:
    - ◊ **Education**: Used as a tool for learning programming and hardware projects.
    - ◊ **Community Projects**: Powers local servers, weather stations, and low-cost computing solutions.
  - **Benefits**:
    - ◊ **Cost-Effective**: Offers computing capabilities at a fraction of the cost of traditional computers.
    - ◊ **Versatile**: Supports various operating systems and programming languages.
- **Government-Subsidized Devices**:
  - **Examples**:
    - ◊ **India's Aakash Tablet**: Developed to provide affordable access to computing and internet in educational institutions.
    - ◊ **OLPC's XO Laptop**: Designed specifically for children in developing countries, rugged and power-efficient.

- **Impact**:
  - ◊ **Educational Advancement**: Students gain access to digital textbooks, educational apps, and online resources.
  - ◊ **Digital Inclusion**: Helps bridge the gap for those who cannot afford standard market devices.

» **Subsidized Internet Services**:

- **Public Wi-Fi Hotspots:**
  - ◊ **Implementation**: Installation of Wi-Fi access points in public areas such as parks, markets, and transportation hubs.
- **Examples**:
  - ◊ **Google Station**: Provided free Wi-Fi in railway stations and other public places in countries like India and Mexico.
  - ◊ **LinkNYC**: Replaces old payphones in New York City with kiosks offering free Wi-Fi, phone calls, and device charging.
- **Benefits**:
  - ◊ **Accessibility**: Allows individuals to connect to the internet without the cost of personal data plans.
  - ◊ **Community Engagement**: Encourages use of online services and resources.

» **Subsidized Data Plans**:

- **Initiatives**:
  - ◊ Partnerships between governments and telecom providers to offer discounted internet packages.
  - ◊ Programs targeting students, low-income families, and rural populations.
- **Impact**:
  - ◊ **Affordability**: Reduces the financial burden of internet access.
  - ◊ **Connectivity**: Enables participation in online education, e-commerce, and communication platforms.

Language and Cultural Relevance

Cultural and linguistic barriers can hinder the effective use of ICT. Ensuring that digital content is available in local languages and is culturally relevant enhances inclusivity and user engagement.

» **Localized Educational Content**:

- **Creation of Native Language Resources**:
  - ◊ Development of educational materials, software, and applications in local languages.

- ◊ Collaboration with local educators and linguists to ensure accuracy and relevance.

- **Examples**:
  - ◊ **Wikipedia in Local Languages**: Supports content creation in over 300 languages, allowing communities to access information in their mother tongue.
  - ◊ **Localized Software**: Open-source projects like LibreOffice offer interfaces in multiple languages.
- **Benefits**:
  - ◊ **Inclusivity**: Makes technology accessible to non-English speakers and those with limited proficiency in global languages.
  - ◊ **Preservation of Cultural Identity**: Supports the use and preservation of local languages and traditions.

» **Translation Tools and Multilingual Platforms**:

- **Translation Technologies**:
  - ◊ **Machine Translation**: Tools like Google Translate facilitate real-time translation of text and speech.
  - ◊ **Speech Recognition**: Voice-activated assistants and dictation software that understand and process local languages.
- **Multilingual Educational Platforms**:
  - ◊ E-learning platforms offering courses and materials in various languages.
  - ◊ Subtitles and dubbing for educational videos and tutorials.
- **Impact**:
  - ◊ **Expanded Access**: Enables more people to benefit from global educational resources.
  - ◊ **Cultural Exchange**: Promotes cross-cultural understanding and collaboration.

Overall Impact of ICT Initiatives

Bridging the digital divide through these initiatives has far-reaching consequences:

» **Economic Empowerment**:

- **Job Creation**: Digital literacy opens up employment opportunities in ICT sectors and beyond.
- **Entrepreneurship**: Access to online markets and resources enables small businesses and startups.

» **Educational Advancement**:

- **Improved Learning Outcomes**: Access to diverse educational materials enhances knowledge and skills.

- **Lifelong Learning**: Encourages continuous personal and professional development.

» **Social Inclusion**:

- **Community Development**: ICT facilitates community organization, advocacy, and participation in governance.
- **Access to Services**: Simplifies access to healthcare, legal aid, and other essential services through e-governance platforms.

» **Innovation and Creativity**:

- **Local Solutions**: Communities can develop technology-based solutions tailored to their specific needs.
- **Cultural Expression**: Digital platforms allow for the sharing and preservation of cultural heritage.

Challenges and Future Directions

While significant progress has been made, challenges persist in fully bridging the digital divide:

» **Infrastructure Deficits**:

- **Connectivity**: Remote areas may still lack reliable internet access due to geographical and economic constraints.
- **Electricity**: Inconsistent power supply hampers the use of ICT devices.

» **Sustainability**:

- **Funding**: Maintaining programs and facilities requires ongoing financial support.
- **Upkeep and Upgrades**: Technology becomes obsolete; continuous updates are necessary.

» **Skill Gaps**:

- **Digital Literacy**: Beyond access, users need training to effectively utilize ICT.
- **Educator Training**: Teachers and trainers require ongoing professional development to stay current.

» **Cultural Resistance**:

- **Acceptance**: Some communities may be hesitant to adopt new technologies due to cultural or generational differences.
- **Relevance**: Ensuring that ICT solutions are tailored to meet the specific needs and contexts of different communities.

Strategies for Effective Implementation

To address these challenges and enhance the impact of ICT initiatives:

» **Public-Private Partnerships**:

- Collaboration between governments, businesses, and NGOs can pool resources and expertise.

» **Community Engagement**:
  - Involving local leaders and members in planning and implementation ensures relevance and acceptance.

» **Policy Frameworks**:
  - Developing supportive policies that promote investment in ICT infrastructure and education.

» **Monitoring and Evaluation**:
  - Regular assessment of programs to measure impact, identify gaps, and inform improvements.

5. Promoting Lifelong Learning

In an era of rapid technological advancement and global connectivity, the concept of lifelong learning has become indispensable. Lifelong learning involves acquiring knowledge and skills continuously throughout an individual's life to adapt to evolving personal, professional, and societal demands. Information and Communication Technology (ICT) serves as a powerful enabler of lifelong learning by providing access to diverse educational resources, fostering skill development, and promoting adaptability in an ever-changing world.

Skill Development Programs

ICT-based skill development programs are designed to equip individuals with competencies that align with the demands of local and global job markets. These programs provide targeted training in technical, vocational, and creative skills, empowering learners to enhance their employability and productivity.

1. Vocational Training

» **Overview**: ICT-driven vocational training focuses on practical, job-oriented skills tailored to industry needs. These programs bridge the gap between education and employment, particularly in regions with high unemployment or underemployment rates.

» **Applications**:
  - Training in sectors like healthcare, construction, agriculture, and IT using simulation tools and online platforms.
  - Certification programs that validate skills for specific roles, such as electricians, machine operators, or digital marketers.

» **Examples**:
  - **Coursera for Campus**: Provides tailored vocational courses for educational institutions and job seekers.
  - **Skill India Mission**: Offers ICT-enabled training in various trades to enhance the employability of India's youth.

- **Impact:**
  - Equips individuals with industry-relevant skills, increasing their chances of securing stable employment.
  - Supports local economies by fostering skilled labor in critical sectors.

2. Digital Platforms for Skill Development

- **Applications:**
  - **Udemy** and **Skillshare:** Offer courses in technical areas like coding, graphic design, digital marketing, and project management, as well as creative fields like photography and writing.
  - **Khan Academy:** Provides foundational skill-building resources in mathematics, science, and finance.
- **Benefits:**
  - **Accessibility:** Courses are available online, allowing learners to access them from any location.
  - **Flexibility:** Programs are self-paced, enabling learners to balance education with other commitments.
  - **Certification:** Many platforms offer certificates that enhance credibility in professional settings.

Adult Education

Adult education is critical for empowering individuals who missed formal education opportunities earlier in life. ICT has transformed adult learning by introducing flexible, accessible, and engaging methods to address literacy, numeracy, and vocational training needs.

1. Mobile Learning Apps and Interactive Tutorials

- **Applications:**
  - Mobile apps deliver bite-sized lessons that cater to adult learners with limited time or resources.
  - Interactive tutorials use multimedia elements like videos, animations, and quizzes to simplify complex concepts.
- **Examples:**
  - **Duolingo:** Helps adults learn new languages in an interactive and gamified environment.
  - **LiteracyPlanet:** Focuses on improving literacy through engaging activities and assessments.
- **Impact:**
  - Enhances basic literacy and numeracy skills, fostering independence and confidence.
  - Promotes self-directed learning, enabling adults to pursue personal and professional growth.

2. Community Radio and Television Broadcasts

- **Applications**:
  - Community radio stations disseminate educational content on topics such as health, agriculture, and civic rights, targeting populations with limited internet access.
  - Educational television programs deliver structured lessons to reach a broad audience.
- **Examples**:
  - **BBC Bitesize**: Offers educational content for learners of all ages through television broadcasts.
  - **Gyan Vani** (India): A dedicated educational radio channel providing information on topics ranging from health to vocational skills.
- **Impact**:
  - Extends educational opportunities to rural and remote areas where digital infrastructure is lacking.
  - Encourages community participation and engagement in lifelong learning initiatives.

Reskilling and Upskilling

The rapid pace of technological innovation and industry transformation has heightened the need for reskilling (learning new skills) and upskilling (enhancing existing skills). ICT enables professionals to stay relevant in their fields and adapt to changing job requirements.

1. Online Certifications and Training Programs

- **Applications**:
  - Specialized courses in emerging fields like artificial intelligence, blockchain, cybersecurity, and data science.
  - Training in soft skills such as communication, leadership, and emotional intelligence, which are increasingly valued in the workplace.
- **Examples**:
  - **LinkedIn Learning**: Provides courses on professional skills with certifications to enhance employability.
  - **AWS Training and Certification**: Offers cloud computing expertise to meet growing demand in tech industries.
- **Benefits**:
  - Supports career progression by equipping professionals with high-demand skills.
  - Offers globally recognized certifications that improve job prospects.

2. Massive Open Online Courses (MOOCs)

- **Overview**: MOOCs are online courses that cater to a large number of participants, providing access to quality education from leading universities and institutions.

» **Examples**:
  - **edX**: Features courses from institutions like Harvard, MIT, and Berkeley, covering a wide range of disciplines.
  - **FutureLearn**: Focuses on skills like coding, business management, and healthcare training.

» **Benefits**:
  - Scalability: Enables millions of learners to access content simultaneously.
  - Diversity: Offers courses ranging from technical to creative disciplines.
  - Cost-Effectiveness: Many MOOCs are free or available at a fraction of the cost of traditional education.

» **Impact**:
  - Democratizes access to world-class education, breaking down socio-economic barriers.
  - Encourages a culture of continuous learning, essential in dynamic industries.

Impact of ICT in Promoting Lifelong Learning

ICT has a profound impact on fostering lifelong learning, benefiting individuals, communities, and economies:

1. **For Individuals**:
   - Empowers personal growth by providing access to diverse knowledge and skills.
   - Enhances employability, enabling individuals to adapt to new roles and industries.
2. **For Communities:**
   - Strengthens local economies by equipping residents with marketable skills.
   - Promotes social inclusion by addressing educational disparities.
3. **For Economies:**
   - Supports workforce transformation in response to technological advancements.
   - Drives innovation and competitiveness by fostering a skilled and adaptable labor force.

6. Driving Social and Economic Development

The influence of Information and Communication Technology (ICT) in community education extends well beyond personal benefits, contributing significantly to broader social and economic development. By equipping individuals with the tools, skills, and knowledge to thrive, ICT transforms communities into hubs of innovation, inclusivity, and progress. Its impact is particularly evident in economic empowerment, social inclusion, and health and well-being.

Economic Empowerment

Economic empowerment is a cornerstone of ICT-driven community education, as it equips individuals with skills and tools that enhance employability, foster entrepreneurship, and drive innovation.

1. Improving Employability Through ICT-Based Education

ICT-enabled education provides individuals with marketable skills that align with current job market demands, increasing their chances of securing stable employment.

» **Skills Training**:
- ICT integrates vocational training in areas such as IT support, coding, graphic design, and digital marketing.
- Programs emphasize both technical skills (e.g., proficiency in software tools) and soft skills (e.g., communication, teamwork).

» **Examples**:
- **Digital Skills for Africa**: A Google initiative offering free training in digital marketing and IT skills to help individuals secure jobs or start businesses.
- **Coursera Career Academy**: Focuses on job-ready skills through certification courses in high-demand fields.

» **Impact**:
- Enhances workforce readiness by closing the skills gap in industries like technology, healthcare, and finance.
- Empowers marginalized groups, including women and youth, to participate in the formal economy.

2. Fostering Entrepreneurship Through ICT

Entrepreneurship is a powerful driver of economic growth, and ICT provides aspiring entrepreneurs with the resources, platforms, and knowledge to succeed.

» **ICT-Driven Tools**:
- **E-commerce Platforms**: Services like Shopify and Etsy enable entrepreneurs to reach global markets.
- **Digital Payment Solutions**: Mobile money services like M-Pesa facilitate secure and convenient financial transactions.
- **Cloud-Based Business Tools**: Platforms like QuickBooks and Zoho streamline operations, from accounting to customer management.

» **Training Programs**:
- ICT-based entrepreneurship courses teach business fundamentals, such as financial planning, marketing, and digital branding.
- Mentorship programs connect entrepreneurs with experienced professionals for guidance and support.

» **Impact**:
- Promotes self-reliance and innovation by enabling individuals to start and grow businesses.

- Stimulates local economies by creating jobs and generating wealth.

Social Inclusion

ICT plays a pivotal role in fostering social inclusion by addressing systemic barriers and empowering underrepresented groups to access education, resources, and opportunities.

1. Addressing Barriers to Education

ICT democratizes access to education for marginalized populations, including women, minorities, and individuals with disabilities.

» **For Women**:
  - ICT-based programs provide flexible learning opportunities that accommodate household responsibilities and cultural norms.
  - Initiatives like **SheCodes** offer coding boot camps designed to empower women in technology.

» **For Minorities and Marginalized Groups**:
  - Community ICT centers in underserved areas provide access to educational resources and skill-building programs.
  - Digital literacy campaigns help bridge the gap between urban and rural communities.

» **Impact**:
  - Increases representation of diverse groups in education and the workforce.
  - Reduces inequality by providing equitable access to opportunities.

2. Amplifying Community Voices Through Digital Storytelling

Digital storytelling platforms enable communities to share their narratives, preserve cultural heritage, and promote social cohesion.

» **Applications**:
  - **Cultural Preservation:**
    - ICT tools like audio and video recording devices help document traditional music, folklore, and oral histories.
  - **Advocacy:**
    - Social media platforms amplify grassroots campaigns, raising awareness of social issues and mobilizing support.

» **Examples**:
  - **StoryCorps**: Collects and shares personal stories to foster understanding and empathy among diverse populations.
  - **YouTube Creators for Change**: Supports creators using the platform to promote social impact and inclusivity.

- **Impact**:
  - Strengthens community identity and pride.
  - Encourages cross-cultural understanding and collaboration.

Health and Well-Being

ICT's role in health and well-being extends to raising awareness, improving access to healthcare services, and empowering communities with knowledge to make informed decisions.

1. ICT-Enabled Health Education

ICT delivers critical health education programs that address issues like nutrition, hygiene, disease prevention, and mental health.

- **Applications:**
  - **Mobile Health Apps:**
    - ◊ Apps like HealthifyMe provide personalized diet plans, fitness tracking, and wellness tips.
  - **Digital Campaigns:**
    - ◊ Social media and messaging apps are used to disseminate public health information, such as COVID-19 prevention guidelines.
- **Impact**:
  - Reduces the spread of preventable diseases by promoting healthy practices.
  - Enhances community awareness of health issues, enabling proactive healthcare decisions.

2. Telemedicine and Remote Healthcare Services

Telemedicine platforms leverage ICT to connect patients with healthcare providers, overcoming barriers like geographical distance and resource constraints.

- **Features**:
  - **Virtual Consultations**: Video conferencing tools enable doctors to assess and treat patients remotely.
  - **Remote Monitoring**: Wearable devices track vital signs and transmit data to healthcare providers in real time.
  - **E-Prescriptions**: Digital platforms streamline the prescription process, reducing errors and improving patient outcomes.
- **Examples**:
  - **Practo**: An online healthcare platform offering teleconsultations, appointment scheduling, and health records management.
  - **Babylon Health**: Uses AI-powered chatbots and video consultations to deliver healthcare services globally.

» **Impact**:
  - Expands access to medical care for underserved populations, including rural communities.
  - Reduces healthcare costs by minimizing the need for physical infrastructure and travel.

The Broader Impact of ICT on Social and Economic Development

ICT's role in driving social and economic development has transformative effects on communities and societies:

1. **Economic Growth**:
   - Enhances productivity and innovation across industries.
   - Supports the creation of new markets and job opportunities.
2. **Social Equity:**
   - Bridges gaps in education, healthcare, and employment, fostering a more inclusive society.
   - Empowers underrepresented groups to contribute to and benefit from societal progress.
3. **Community Resilience:**
   - Strengthens communities by providing tools for disaster preparedness, response, and recovery.
   - Encourages civic engagement through e-governance platforms and digital participation.

Chapter 2

IT Enabled Services (ITES)

Information Technology Enabled Services (ITES) refers to a broad range of technology-driven services that leverage IT infrastructure to deliver value-added solutions to businesses and end-users. These services play a critical role in improving efficiency, enhancing customer satisfaction, and enabling data-driven decision-making across industries. Among the most significant ITES are **call centers**, **helpdesks**, and **data warehouses**, each of which fulfills distinct but interconnected functions in today's digital landscape.

1. Call Centers

Call centers are essential facilities for managing customer interactions, acting as the frontline for businesses to handle communication effectively. They are pivotal in maintaining robust **Customer Relationship Management (CRM)** strategies, enabling companies to build long-term relationships with their customers while ensuring operational efficiency. By addressing a high volume of inbound and outbound communications, call centers serve as a critical touchpoint for resolving issues, enhancing customer satisfaction, and driving business growth.

Key Features of Call Centers

Call centers have evolved significantly, integrating advanced technologies and multi-channel communication systems to provide comprehensive services. These features enable businesses to meet customer expectations effectively, enhancing both responsiveness and overall service quality.

1. Inbound Call Centers

Inbound call centers focus on managing incoming calls from customers, often related to support, inquiries, or service-related issues.

- **Functionality**:
    - Designed to address customer concerns, these centers are reactive in nature, responding to inquiries, complaints, or technical issues.
    - They serve as a primary point of contact for businesses, ensuring customers have a reliable channel for communication.
- **Examples**:
    - **Telecom Providers**: Companies like Verizon or Airtel manage billing inquiries, technical support, and service upgrades.

DOI: 10.4324/9781003750444-2

    - **E-commerce Platforms**: Businesses like Amazon and Flipkart handle product returns, order tracking, and delivery-related concerns.
- **Key Features**:
    - Toll-free numbers or dedicated lines for customer queries.
    - Integration with CRM systems to access customer history and provide personalized support.
    - Automated voice response (IVR) systems to guide callers efficiently.
- **Impact**:
    - Increases customer retention by resolving issues promptly.
    - Strengthens trust by providing reliable support for product or service-related concerns.

2. Outbound Call Centers

Outbound call centers are designed to proactively reach out to customers for purposes such as sales, feedback collection, or payment reminders.

- **Functionality**:
    - These centers focus on lead generation, customer outreach, and promoting products or services.
    - They also play a key role in gathering valuable insights through surveys and feedback campaigns.
- **Examples**:
    - **Insurance Companies**: Providers like AIG or LIC contact customers for policy renewals, upselling additional coverage, or cross-selling related services.
    - **Retail Chains**: Businesses like Walmart engage customers with exclusive offers or loyalty program updates.
- **Key Features**:
    - Dialer systems that automate the process of contacting potential customers.
    - Analytics-driven targeting for effective and personalized communication.
    - Detailed call scripts to ensure consistency and compliance with regulations.
- **Impact**:
    - Drives revenue growth through targeted marketing and sales initiatives.
    - Builds stronger customer relationships through consistent follow-ups and tailored communication.

3. Omnichannel Support

Modern call centers are no longer limited to voice communication. The integration of multiple communication channels has transformed call centers into comprehensive customer support hubs.

- **Functionality:**
    - Omnichannel call centers enable seamless transitions between communication modes such as phone calls, emails, live chats, social media, and messaging platforms.
    - Customers can interact through their preferred channels, enhancing convenience and accessibility.
- **Examples:**
    - A customer initiating a complaint on Twitter receives a detailed resolution through email and a follow-up via phone.
- **Key Features:**
    - Unified communication platforms that consolidate interactions across all channels.
    - Consistent customer experience with centralized data access for agents.
    - Real-time updates to ensure continuity of service across channels.
- **Impact:**
    - Enhances customer satisfaction by offering flexibility and personalized support.
    - Improves operational efficiency by consolidating interaction histories.

4. Cloud-Based Call Centers

The adoption of cloud technology has revolutionized the call center industry, enabling greater flexibility, scalability, and cost-effectiveness.

- **Functionality:**
    - Cloud-based call centers operate on virtual infrastructure, allowing agents to work remotely and organizations to scale operations as needed.
    - They eliminate the need for on-premises hardware, reducing setup costs and maintenance.
- **Examples:**
    - Virtual call centers supporting global businesses like Uber or Airbnb, where agents handle customer inquiries from various locations.
- **Key Features:**
    - Easy deployment with minimal hardware requirements.
    - Enhanced scalability to manage seasonal spikes in customer demand.
    - Secure, cloud-based data storage with robust encryption protocols.

- **Impact**:
    - Reduces operational costs while maintaining service quality.
    - Offers scalability for small businesses and startups, enabling them to compete with larger enterprises.

Applications of Call Centers

Call centers play a versatile role across industries, supporting various business functions. Their ability to manage communication at scale makes them indispensable for ensuring customer satisfaction and operational efficiency.

1. Customer Support

Providing timely and effective support is one of the primary applications of call centers.

- **Functionality**:
    - Addressing customer queries, resolving technical issues, and handling complaints.
    - Offering real-time assistance to enhance the customer experience.
- **Examples**:
    - **Amazon**: Its customer service centers resolve issues related to delayed deliveries, product returns, and account management.
    - **Tech Companies**: Brands like Dell and Apple provide technical support for troubleshooting hardware and software problems.
- **Impact**:
    - Builds customer trust by ensuring reliable support.
    - Enhances brand reputation through exceptional service delivery.

2. Telemarketing

Telemarketing involves promoting products or services directly to potential customers through outbound calls.

- **Functionality**:
    - Engaging prospects with personalized offers, discounts, or product demonstrations.
    - Generating leads and following up with potential clients to convert sales.
- **Examples**:
    - **Telecom Companies**: Providers like AT&T contact customers to offer promotional packages or upgrades.
    - **E-Learning Platforms**: Platforms like Coursera or Skillshare engage users with trial offers or subscription plans.

- **Impact**:
    - Drives revenue growth through direct sales and upselling.
    - Increases market reach by targeting specific customer segments.

3. Collections

Call centers play a vital role in managing overdue payments and debt recovery.

- **Functionality**:
    - Reminding customers of payment deadlines or assisting them with repayment plans.
    - Managing collections for loans, credit cards, or subscription-based services.
- **Examples**:
    - Banks like Chase or Citibank contact customers for EMI reminders and overdue payments.
    - Subscription services like Netflix send reminders for expiring memberships.
- **Impact**:
    - Improves cash flow by reducing overdue balances.
    - Maintains customer relationships through empathetic and solution-oriented communication.

4. Market Research

Call centers are essential for gathering customer feedback and conducting surveys.

- **Functionality**:
    - Collecting data on customer preferences, satisfaction levels, and market trends.
    - Conducting post-purchase surveys to assess product or service quality.
- **Examples**:
    - Market research firms use call centers to gauge customer sentiment on new product launches.
    - Companies like Nielsen conduct telephonic surveys to understand consumer behavior.
- **Impact**:
    - Provides actionable insights for strategic decision-making.
    - Enhances product development by aligning with customer needs.

Benefits of Call Centers

Call centers offer numerous advantages that contribute to customer satisfaction, business growth, and operational excellence.

1. Enhanced Customer Satisfaction

- Ensures 24/7 support to address customer needs promptly.
- Builds trust and loyalty by providing personalized and empathetic service.

2. Operational Efficiency

- Centralizes communication processes, minimizing redundancies and streamlining workflows.
- Reduces response times through automated tools like IVR systems.

3. Cost Savings

- Outsourcing call centers to countries with lower labor costs significantly reduces operational expenses.
- Cloud-based solutions eliminate the need for expensive hardware and maintenance.

4. Business Growth

- Drives sales through targeted telemarketing and lead generation campaigns.
- Expands market reach by connecting with diverse customer segments.

Challenges in Call Centers

Despite their benefits, call centers face challenges that can affect their efficiency and effectiveness.

1. High Turnover Rates

- Stressful work environments and repetitive tasks lead to employee attrition.
- Addressing turnover requires robust training programs and employee engagement initiatives.

2. Technological Complexity

- Integrating CRM systems, communication platforms, and analytics tools demands continuous upgrades and investments.
- Maintaining data security and ensuring seamless operations can be resource-intensive.

3. Language and Cultural Barriers

- Offshore call centers may struggle with understanding and addressing the cultural nuances of customers.
- Training in cultural sensitivity and language proficiency is essential to mitigate these issues.

2. Helpdesks

Helpdesks are a vital component of modern IT infrastructure, providing technical support and assistance to users of IT systems and services. They act as a centralized point of contact for resolving technical issues, offering guidance, and maintaining operational continuity in organizations. Helpdesks cater to both internal users, such as employees, and external users, such as customers, ensuring seamless access to IT services.

Key Features of Helpdesks

Helpdesks are designed with specific features to efficiently handle support requests and deliver effective solutions. These features enable helpdesk teams to streamline workflows, improve response times, and provide consistent support.

1. Ticketing Systems

- **Overview:**
  - Ticketing systems are core to helpdesk operations, allowing teams to log, categorize, and prioritize user requests systematically.
- **Functionality:**
  - Automatically assigns unique identifiers (tickets) to each support request for tracking.
  - Categorizes issues based on type, urgency, and complexity.
  - Assigns tickets to appropriate teams or agents based on expertise.
- **Benefits:**
  - Enhances organization by creating a structured workflow for issue resolution.
  - Reduces redundancy by avoiding duplication of efforts on similar issues.
  - Provides a clear record of resolved and unresolved tickets, aiding in performance tracking.

2. Knowledge Bases

- **Overview:**
  - Knowledge bases are repositories of information designed to empower users to resolve common issues independently.
- **Functionality:**
  - Includes articles, FAQs, video tutorials, and troubleshooting guides.
  - Organizes content by categories or search functionality for easy access.
  - Continuously updated with solutions to recurring issues.
- **Benefits:**
  - Empowers users with self-service options, reducing reliance on helpdesk agents.
  - Minimizes response times for routine queries.
  - Enhances knowledge sharing within organizations.

3. Multichannel Support

- **Overview:**
  - Helpdesks provide support through various communication channels, ensuring accessibility for users with different preferences.

- **Functionality**:
  - Includes phone, email, live chat, online portals, and mobile apps.
  - Integrates with CRM systems to provide consistent support across channels.
  - Offers real-time assistance through chatbots or live agents.
- **Benefits**:
  - Increases user convenience by offering multiple options for contacting support.
  - Improves efficiency by centralizing interaction history for all channels.
  - Enhances customer satisfaction with timely and personalized assistance.

4. Automation

- **Overview**:
  - Automation tools enhance helpdesk efficiency by handling routine tasks and freeing up agents for more complex issues.
- **Functionality**:
  - AI-powered chatbots address frequently asked questions and guide users through troubleshooting steps.
  - Automated ticket assignment prioritizes urgent issues and routes them to appropriate agents.
  - Predictive analytics identify potential issues before they escalate.
- **Benefits**:
  - Reduces response times for repetitive queries.
  - Optimizes resource allocation by minimizing manual intervention.
  - Improves accuracy in issue categorization and resolution.

Applications of Helpdesks

Helpdesks serve diverse applications across industries, addressing technical, customer, and operational needs. Their versatility ensures seamless IT operations and enhances user experience.

1. IT Support

- **Functionality**:
  - Provides technical assistance for hardware, software, and network-related issues.
  - Supports employees with tools and resources to perform their roles effectively.
- **Examples**:
  - Internal IT helpdesks resolving problems with employee laptops, email systems, or software installations.

- IT teams managing password resets and account recovery requests.

- **Impact**:
  - Ensures uninterrupted workflow for employees by minimizing technical disruptions.
  - Reduces downtime, enhancing productivity.

2. Customer Support

- **Functionality**:
  - Assists customers with product setup, troubleshooting, and maintenance.
  - Addresses post-purchase concerns, such as warranty claims or usage guidance.
- **Examples**:
  - Technical support for software like Microsoft Office or Adobe Creative Suite.
  - Hardware companies like Dell or HP helping users configure devices or resolve malfunctions.
- **Impact**:
  - Enhances customer satisfaction and trust in the product or service.
  - Encourages repeat business and brand loyalty.

3. Incident Management

- **Functionality**:
  - Identifies, addresses, and resolves technical incidents to minimize downtime and operational disruptions.
- **Examples**:
  - Telecom companies resolving network outages or slow connectivity reported by users.
  - Cloud service providers managing server outages affecting hosted applications.
- **Impact**:
  - Maintains service continuity and reliability.
  - Reduces the financial and reputational impact of technical failures.

4. Knowledge Sharing

- **Functionality**:
  - Provides users with self-help resources for resolving minor issues independently.
  - Facilitates knowledge transfer within organizations for consistent support.

- **Examples**:
  - FAQs and troubleshooting guides hosted on company websites.
  - Video tutorials explaining software installation or feature usage.
- **Impact**:
  - Reduces the volume of support requests by enabling self-service.
  - Enhances user confidence in handling common issues.

Benefits of Helpdesks

Helpdesks offer significant advantages for organizations and users, contributing to operational efficiency, cost savings, and user satisfaction.

1. Improved Efficiency

- How:
  - Streamlines issue resolution with ticketing systems and prioritization protocols.
  - Automates repetitive tasks, allowing agents to focus on complex issues.
- **Impact**:
  - Enhances workflow organization and reduces resolution times.
  - Ensures consistent service quality across all support requests.

2. Customer Empowerment

- How:
  - Self-service options like knowledge bases and FAQs allow users to resolve issues independently.
  - Multichannel support ensures accessibility and convenience.
- **Impact**:
  - Reduces frustration by minimizing wait times.
  - Encourages users to take proactive steps in troubleshooting.

3. Cost Savings

- How:
  - Automation and self-service options reduce reliance on human agents for routine queries.
  - Centralized systems optimize resource utilization.
- **Impact**:
  - Lowers operational costs without compromising service quality.

- Increases scalability by managing higher volumes of requests with minimal additional resources.

4. Data Insights

- **How:**
  - Analytics tools track common issues, resolution times, and user feedback.
  - Predictive analytics identify patterns to prevent recurring problems.
- **Impact:**
  - Enables proactive IT management by addressing potential issues early.
  - Enhances decision-making with actionable insights.

Challenges in Helpdesks

Despite their benefits, helpdesks face challenges that can hinder their effectiveness and user satisfaction.

1. Scalability

- **Challenge:**
  - Increasing volumes of support requests can overwhelm small helpdesk teams.
- **Solution:**
  - Implement automation tools and AI-driven chatbots to handle routine queries.
  - Expand team size or outsource services during peak periods.

2. Response Times

- **Challenge:**
  - Delayed responses lead to dissatisfaction among users, particularly for critical issues.
- **Solution:**
  - Introduce tiered support systems to prioritize urgent requests.
  - Use analytics to monitor and improve average resolution times.

3. Technical Expertise

- **Challenge:**
  - Helpdesk staff require ongoing training to keep up with evolving technologies.
- **Solution:**
  - Conduct regular training sessions and certifications for helpdesk teams.
  - Encourage knowledge-sharing initiatives within the organization.

3. Data Warehouses

Data warehouses are highly specialized systems designed for the efficient storage, integration, and analysis of large volumes of structured and unstructured data. As an essential component of modern Information and Communication Technology (ICT), data warehouses form the foundation for business intelligence (BI) and analytics, enabling organizations to make informed, data-driven decisions. By centralizing data from multiple sources and optimizing it for analysis, data warehouses help organizations uncover trends, forecast outcomes, and gain a competitive edge.

Key Features of Data Warehouses

Data warehouses are distinguished by their robust architecture and ability to handle complex data operations. These features ensure the seamless storage, integration, and analysis of diverse datasets.

1. Centralized Data Storage

- **Overview**:
  - Data warehouses act as a central repository that consolidates data from multiple sources, including Customer Relationship Management (CRM) systems, Enterprise Resource Planning (ERP) systems, and transactional databases.
- **Functionality**:
  - Collects data from disparate systems into a unified storage location.
  - Organizes data into subject-specific areas such as sales, marketing, finance, and operations.
  - Supports historical data storage for trend analysis and forecasting.
- **Benefits**:
  - Reduces data silos, enabling seamless cross-departmental analysis.
  - Enhances data accessibility by providing a single source of truth.

2. Data Integration

- **Overview**:
  - Combines data from different formats, structures, and systems into a consistent, unified format that can be analyzed effectively.
- **Functionality**:
  - Extract, Transform, Load (ETL) processes ensure data consistency by cleansing, normalizing, and integrating data from diverse sources.
  - Handles structured data (e.g., tables, spreadsheets) and unstructured data (e.g., social media posts, images, emails).
- **Benefits**:
  - Ensures data accuracy and reliability for downstream analytics.

  - Enables comprehensive insights by merging data from multiple domains.

3. Scalability

- **Overview**:
  - Data warehouses are designed to handle increasing volumes of data as organizations grow.
- **Functionality**:
  - Scalable architectures, including cloud-based solutions, allow for dynamic resource allocation.
  - Supports high-performance operations as data volumes and complexity increase.
- **Benefits**:
  - Accommodates future data growth without disrupting existing operations.
  - Reduces the need for frequent infrastructure overhauls.

4. High Performance

- **Overview**:
  - Optimized for complex analytical operations and fast query processing, data warehouses provide high-speed insights.
- **Functionality**:
  - Leverages indexing, partitioning, and caching to accelerate query performance.
  - Uses columnar storage to optimize analytical queries.
- **Benefits**:
  - Reduces time-to-insight, enabling faster decision-making.
  - Handles concurrent queries from multiple users without performance degradation.

Applications of Data Warehouses

Data warehouses are integral to various industries, supporting diverse applications that leverage data for strategic, operational, and tactical purposes.

1. Business Intelligence and Analytics

- **Functionality**:
  - Enables real-time and predictive analytics, empowering organizations to make strategic decisions based on comprehensive data insights.
- **Examples**:
  - **Retailers**: Analyze customer purchase patterns to optimize inventory and personalize marketing campaigns.

  - **Financial Services**: Track investment trends to identify profitable opportunities.
- **Impact**:
  - Enhances market responsiveness by identifying trends and customer preferences.
  - Supports scenario modeling for risk assessment and strategy formulation.

2. Financial Reporting

- **Functionality**:
  - Consolidates financial data from various systems to generate accurate and timely financial statements.
- **Examples**:
  - **Banks**: Aggregate transaction data to monitor revenue streams, track expenses, and evaluate profitability.
  - **Corporations**: Use consolidated financial reports for compliance with regulatory requirements.
- **Impact**:
  - Improves transparency and accountability in financial operations.
  - Facilitates compliance with industry standards and legal mandates.

3. Healthcare Analytics

- **Functionality**:
  - Integrates patient records, diagnostic data, and treatment outcomes to enhance healthcare delivery and resource allocation.
- **Examples**:
  - **Hospitals**: Identify trends in disease prevalence to prepare resources and optimize staffing.
  - **Pharmaceutical Companies**: Analyze clinical trial data to expedite drug development.
- **Impact**:
  - Enhances patient care through data-driven decision-making.
  - Reduces costs by streamlining operations and improving resource utilization.

4. Supply Chain Optimization

- **Functionality**:
  - Tracks and analyzes data across the supply chain to identify inefficiencies and improve operational efficiency.

- **Examples**:
    - **E-Commerce Companies**: Optimize delivery routes using historical shipping data and real-time logistics.
    - **Manufacturers**: Predict demand patterns to minimize inventory costs and reduce stockouts.
- **Impact**:
    - Improves supply chain resilience and adaptability.
    - Enhances customer satisfaction through timely deliveries and consistent service quality.

Benefits of Data Warehouses

Data warehouses provide significant benefits to organizations, enabling them to leverage data for competitive advantage and operational excellence.

1. Enhanced Decision-Making

- **How**:
    - Aggregates and analyzes data from multiple sources to provide actionable insights.
    - Enables real-time and predictive analytics for proactive decision-making.
- **Impact**:
    - Supports strategic planning and operational improvements.
    - Facilitates informed decision-making at all organizational levels.

2. Improved Data Quality

- **How**:
    - Cleans and standardizes data through ETL processes to ensure accuracy and consistency.
    - Eliminates duplicate and irrelevant data for more reliable analysis.
- **Impact**:
    - Enhances trust in data-driven insights.
    - Reduces errors in reporting and analysis.

3. Competitive Advantage

- **How**:
    - Provides advanced analytics to identify market trends and customer preferences.
    - Enables faster adaptation to changing business environments.

- **Impact**:
  - Positions organizations as market leaders through timely and accurate insights.
  - Drives innovation by uncovering new opportunities.

4. Regulatory Compliance

- **How**:
  - Ensures accurate and auditable record-keeping for regulatory requirements.
  - Tracks historical data for legal and compliance reporting.
- **Impact**:
  - Reduces the risk of non-compliance penalties.
  - Enhances transparency in operations.

Current Status of Applications in IT Enabled Services

The rapid evolution of Information Technology Enabled Services (ITES) has profoundly reshaped how organizations operate and deliver value. This transformation, fueled by advancements in technology, changing consumer behavior, and the globalization of services, has made ITES indispensable in sectors such as finance, healthcare, retail, education, and logistics. Key components of ITES—**call centers**, **helpdesks**, and **data warehouses**—play pivotal roles in improving efficiency, customer satisfaction, and decision-making.

1. Call Centers:

Modern call centers have undergone a significant transformation, shifting from simple query resolution centers to comprehensive customer engagement hubs. This evolution has been driven by technological advancements, the need for streamlined operations, and a growing emphasis on customer-centric experiences. Today's call centers focus on delivering personalized, efficient, and proactive service through the integration of advanced tools and techniques, enabling businesses to build lasting customer relationships.

Transformation in Call Centers

Historically, call centers were designed to address specific customer issues via telephone. While this basic function remains integral, the scope of call centers has expanded dramatically. They now serve as critical touchpoints for businesses to enhance customer satisfaction, gather valuable insights, and drive revenue growth. Key drivers of this transformation include:

1. **Technological Innovation**:
   - The adoption of technologies like artificial intelligence (AI), cloud computing, and predictive analytics has redefined how call centers operate.
   - Automation and integration tools enable faster response times and improved accuracy.
2. **Changing Consumer Expectations**:
   - Customers expect prompt, convenient, and personalized support.

   - The growing reliance on digital communication channels has necessitated omnichannel solutions.

3. **Globalization and Remote Work:**
   - The need for scalable, flexible solutions has led to the widespread adoption of cloud-based call centers.
   - Remote work capabilities have become essential, particularly in response to events like the COVID-19 pandemic.

Key Developments in Call Centers

Modern call centers incorporate several advanced features and capabilities to address the dynamic needs of businesses and customers.

1. Omnichannel Integration

- **Overview:** Call centers now provide unified support across multiple communication channels, including voice, email, live chat, social media, and messaging platforms. This ensures a seamless experience for customers, regardless of their preferred communication method.
- **Functionality:**
  - Customers can transition between channels without losing context.
  - Integrated systems maintain a centralized record of all interactions, allowing agents to provide consistent support.
- **Benefits:**
  - Enhances customer convenience and satisfaction.
  - Reduces response times and improves operational efficiency.
- **Examples:**
  - A customer might start a complaint via Twitter, receive a detailed follow-up through email, and finalize the resolution with a phone call.
  - Tools like Zendesk and Salesforce enable seamless omnichannel integration.

2. Cloud-Based Call Centers

- **Overview:** Cloud technology has revolutionized the call center industry by enabling remote operations, scalability, and cost efficiency. Cloud-based solutions provide businesses with flexibility and adaptability in a rapidly changing environment.
- **Benefits:**
  - **Flexibility:** Agents can work from any location, ensuring operational continuity even during disruptions like natural disasters or global pandemics.
  - **Scalability:** Businesses can easily scale their operations to accommodate fluctuations in call volumes, such as seasonal peaks.

- **Examples**:
  - **Amazon Connect**: A cloud-based contact center offering features like AI-driven insights, real-time monitoring, and analytics dashboards.
  - **RingCentral**: A unified communication platform that integrates voice, video, and messaging.
- **Impact**:
  - Reduces infrastructure costs by eliminating the need for physical hardware.
  - Enhances data security through cloud-based encryption and compliance measures.

3. Artificial Intelligence (AI) and Automation

- **Overview**: AI and automation tools have transformed call center operations, enabling faster resolution of routine queries and improving the overall efficiency of human agents.
- **Applications**:
  - **Chatbots**: AI-powered bots like IBM Watson assist with answering frequently asked questions (FAQs), guiding users through self-service options.
  - **Speech Analytics**: Tools analyze customer sentiment during calls, helping agents adapt their responses and improve service quality.
  - **Predictive Analytics**: Anticipates customer needs based on historical data, enabling proactive engagement.
- **Examples**:
  - AI chatbots can handle queries related to account balances, order statuses, or basic troubleshooting, freeing up agents for more complex tasks.
  - Speech analytics tools like Verint and CallMiner provide actionable insights for agent training and quality assurance.
- **Benefits**:
  - Enhances response times for routine inquiries.
  - Reduces operational costs by automating repetitive tasks.

4. Focus on Personalization

- **Overview**: Advanced Customer Relationship Management (CRM) systems enable call centers to offer personalized support, tailoring interactions to individual customer needs and preferences.
- **Functionality**:
  - Agents have access to detailed customer histories, including past interactions, purchase records, and preferences.
  - Dynamic scripting tools adjust call flows based on customer data.

- **Benefits**:
    - Builds trust and loyalty by addressing customers' unique concerns.
    - Enhances overall customer satisfaction and retention rates.
- **Examples**:
    - Telecom companies offering personalized upgrade recommendations based on usage patterns.
    - E-commerce platforms suggesting complementary products during post-purchase support calls.

Applications of Call Centers

Call centers cater to a wide range of business functions, from customer support to sales and market research. Their versatility makes them indispensable across industries.

1. Customer Support

- **Functionality**:
    - Ensures prompt resolution of customer concerns, such as technical issues, billing disputes, and product-related queries.
    - Provides real-time assistance through voice or digital channels.
- **Examples**:
    - **Telecom Companies**: AT&T and Verizon handle complaints about network outages or billing errors.
    - **E-commerce Giants**: Amazon resolves issues related to delivery tracking, product returns, and refunds.
- **Impact**:
    - Enhances customer loyalty by addressing concerns efficiently.
    - Strengthens brand reputation through exceptional service delivery.

2. Sales and Marketing

- **Functionality**:
    - Outbound call centers focus on generating leads, upselling, and cross-selling.
    - Engage prospects with tailored offers and promotional campaigns.
- **Examples**:
    - **Insurance Companies**: Providers like Allstate and Geico offer customized policies through proactive outreach.
    - **Retail Chains**: Brands like Macy's inform customers about exclusive sales or loyalty programs.

- **Impact**:
  - Drives revenue growth by targeting potential and existing customers.
  - Increases market penetration through personalized marketing efforts.

3. Market Research

- **Functionality**:
  - Conducts surveys to gather customer feedback and insights on market trends.
  - Analyzes data to refine products, services, and marketing strategies.
- **Examples**:
  - Retail companies use call-based surveys to assess customer satisfaction with recent purchases.
  - Market research firms collect opinions on new product launches.
- **Impact**:
  - Provides actionable insights for strategic decision-making.
  - Aligns product offerings with customer needs and preferences.

4. Collections

- **Functionality**:
  - Manages overdue payments and debt recovery through efficient communication strategies.
  - Negotiates repayment plans to maintain customer relationships.
- **Examples**:
  - Financial institutions like Citibank contact customers for EMI collections.
  - Subscription services like Netflix remind users of pending payments.
- **Impact**:
  - Improves cash flow by reducing outstanding balances.
  - Maintains customer goodwill by offering empathetic solutions.

Chapter 3

Government Policy on ICT and Emerging Research

Government policies on Information and Communication Technology (ICT) form the foundation for a nation's digital transformation, fostering innovation, inclusivity, and economic growth. In an era defined by rapid technological advancement, robust ICT policies are vital for addressing societal challenges, creating digital equity, and building a competitive global economy.

Overview of ICT Policy Frameworks

ICT policies encompass a broad spectrum of initiatives designed to integrate technology into key areas of national development, such as education, healthcare, infrastructure, and governance. These policies typically aim to:

1. **Enhance Digital Access and Equity**: Address the digital divide by improving infrastructure and providing affordable access to technology.
2. **Promote Innovation and Research**: Encourage the development of new technologies and support startups and research institutions.
3. **Facilitate E-Governance**: Streamline government processes and ensure transparency through digital tools.
4. **Boost Economic Growth**: Use ICT to create jobs, enhance productivity, and attract investments.
5. **Ensure Cybersecurity and Data Privacy**: Develop frameworks to protect citizens and businesses from digital threats.

India's National ICT Policies:

India serves as a compelling case study for ICT-driven transformation. Over the years, the Indian government has launched a series of policies to leverage ICT for inclusive development and global competitiveness. These policies reflect a commitment to bridging the digital divide, fostering innovation, and creating a robust digital economy.

1. National Policy on Information Technology, 2012

The **National Policy on Information Technology (2012)** was a landmark initiative aimed at transforming India into a global IT leader. The policy outlined a vision to use ICT for inclusive growth, addressing socio-economic challenges, and fostering innovation.

DOI: 10.4324/9781003750444-3

- **Objectives**:
  1. **Enhance Global IT Standing**:
     - Strengthen India's position as a global IT and IT-enabled services (ITES) hub.
     - Encourage exports of IT products and services to generate revenue and create jobs.
  2. **Promote Inclusive Growth**:
     - Ensure ICT reaches underserved areas, promoting digital literacy and access to technology.
  3. **Foster Innovation**:
     - Support research and development in emerging technologies such as cloud computing, artificial intelligence, and mobile applications.
- **Implementation Highlights**:
  - Development of software technology parks and special economic zones (SEZs) to attract IT companies.
  - Public-private partnerships (PPPs) to accelerate technology adoption in rural areas.
  - Incentives for startups and small and medium enterprises (SMEs) in the IT sector.
- **Impact**:
  - Contributed significantly to the IT sector's GDP share.
  - Boosted employment opportunities, particularly for youth and women in ITES.

2. Digital India Initiative (2015)

The **Digital India Initiative**, launched in 2015, represents one of the most ambitious ICT policies globally. Its aim is to transform India into a digitally empowered society and knowledge economy.

- **Core Components**:
  1. **Digital Infrastructure as a Utility for Every Citizen**:
     - Ensure universal access to high-speed internet.
     - Build digital literacy and improve mobile connectivity in rural and remote areas.
  2. **Governance and Services on Demand**:
     - Digitize government services to make them accessible, efficient, and transparent.
     - Develop e-governance platforms for services like tax filing, land records, and health schemes.

3. **Digital Empowerment of Citizens:**
    - Promote digital literacy through initiatives such as PMGDISHA (Pradhan Mantri Gramin Digital Saksharta Abhiyan).
    - Enable citizens to securely access government services online.

- **Key Projects**:
    - **Aadhaar**: A biometric-based unique identification system, now central to various e-governance services.
    - **BharatNet**: A rural broadband initiative to connect all 250,000 Gram Panchayats with high-speed internet.
    - **DigiLocker**: A digital document storage service to securely store and access important documents.
- **Impact**:
    - Enhanced efficiency in service delivery and reduced corruption.
    - Empowered citizens through greater digital inclusion.
    - Boosted the digital economy with initiatives like Digital Payments and e-Commerce.

3. National Digital Communications Policy, 2018

The **National Digital Communications Policy (NDCP)** of 2018 seeks to modernize India's digital infrastructure and position it among the top 50 nations in the ICT Development Index.

- **Objectives**:
    1. Expand digital infrastructure to provide high-speed broadband for all.
    2. Enhance global competitiveness by fostering innovation and entrepreneurship.
    3. Ensure digital sovereignty through data protection and privacy frameworks.
- **Initiatives**:
    - **Broadband for All**: Extend fiber-optic connectivity to every village by 2022.
    - **Next-Gen Technologies**: Encourage the adoption of 5G, IoT (Internet of Things), and AI.
    - **Skill Development**: Train the workforce in advanced ICT skills to meet industry demands.
- **Achievements**:
    - Significant expansion of broadband connectivity.
    - Increased investment in 5G technology and infrastructure.

4. National Policy on ICT in School Education

Education is a cornerstone of national development, and ICT integration into school education is pivotal for preparing youth for a digital future.

- **Goals**:
    1. Equip students with 21st-century skills through the integration of ICT in teaching and learning.
    2. Address barriers to education in underserved areas using digital tools.
    3. Develop e-content in regional languages to promote inclusivity.
- **Implementation Strategies**:
    - Smart classrooms equipped with digital projectors and interactive boards.
    - Teacher training programs to enhance digital teaching skills.
    - Partnerships with ed-tech companies to create high-quality digital content.
- **Impact**:
    - Improved learning outcomes in schools.
    - Enhanced access to quality education in rural areas.

5. Emerging Policies and Future Directions

As technology evolves, governments must adapt their ICT policies to address emerging challenges and opportunities.

1. Artificial Intelligence (AI) Policies

- Governments globally, including India, are investing in AI to enhance productivity and innovation.
- Example: The establishment of AI research labs and funding for AI startups.

2. Cybersecurity and Data Protection

- With growing concerns about data privacy, governments are introducing stringent cybersecurity laws.
- Example: India's Personal Data Protection Bill, designed to regulate data collection and ensure user privacy.

3. Green ICT Policies

- Promote energy-efficient technologies to reduce the carbon footprint of ICT infrastructure.
- Encourage the recycling of e-waste and sustainable practices in the tech industry.

Global Trends in ICT Policy

ICT policies worldwide share common themes of inclusivity, innovation, and governance transparency. Key global examples include:

1. **Singapore's Smart Nation Initiative:**
    - Focus on integrating IoT and AI to enhance urban living and governance.
2. **European Union's Digital Strategy:**
    - Emphasis on data privacy and ethical AI.
3. **South Korea's Digital New Deal:**
    - Investments in 5G, cloud computing, and green technologies.

Emerging Research Issues in Information and Communication Technology (ICT)

The field of Information and Communication Technology (ICT) is rapidly evolving, driven by advancements in technology and its integration into diverse aspects of society. These transformations give rise to critical research areas that seek to address the challenges and opportunities presented by emerging technologies. Research in ICT not only fosters innovation but also ensures the ethical, sustainable, and inclusive growth of technology applications.

1. Artificial Intelligence (AI) and Machine Learning (ML)

AI and ML have become cornerstones of ICT research, revolutionizing fields such as healthcare, education, finance, and logistics. However, their rapid adoption has also raised complex challenges that require in-depth exploration.

Key Research Areas in AI and ML:

1. **Ethical AI:**
    - Ensuring AI systems are transparent, unbiased, and fair.
    - Addressing algorithmic biases that disproportionately affect marginalized groups.
    - Developing frameworks for accountability in AI decision making processes.
2. **Explainable AI (XAI):**
    - Creating models that provide clear and interpretable outputs.
    - Enhancing user trust by making AI decisions more understandable.
    - Applications: Medical diagnosis systems, autonomous vehicles, and legal AI tools.
3. **AI for Social Good:**
    - Leveraging AI to address societal challenges such as poverty, education inequality, and climate change.
    - Example: AI models predicting natural disasters to optimize disaster management efforts.
4. **Autonomous Systems:**
    - Enhancing the reliability and safety of autonomous vehicles, drones, and robots.
    - Developing standards for human-machine collaboration in industrial settings.

2. Cybersecurity and Privacy

As digital transformation accelerates, protecting sensitive information and infrastructure has become a top priority. Research in cybersecurity addresses the complexities of securing systems in an interconnected world.

Key Research Areas in Cybersecurity:

1. **Threat Detection and Response**:
   - Developing advanced intrusion detection systems (IDS) using AI and ML.
   - Proactive threat hunting to identify vulnerabilities before exploitation.
2. **Quantum Cryptography**:
   - Preparing encryption methods to withstand quantum computing attacks.
   - Research into post-quantum cryptography for future-proof security systems.
3. **IoT Security**:
   - Addressing vulnerabilities in Internet of Things (IoT) devices.
   - Creating lightweight encryption techniques suitable for resource-constrained devices.
4. **User Privacy**:
   - Ensuring compliance with data protection regulations such as GDPR and CCPA.
   - Enhancing privacy-preserving techniques like differential privacy and homomorphic encryption.
5. **Zero-Trust Architecture**:
   - Advancing frameworks that assume no system element is inherently secure.
   - Implementing multi-layered security measures to protect critical assets.

3. Data Ethics and Governance

With the exponential growth of data generation, issues surrounding its ethical use and governance have taken center stage. Research in this domain focuses on creating robust frameworks for managing data responsibly.

Key Research Areas in Data Ethics:

1. **Data Sovereignty**:
   - Ensuring that data collected in a region is governed by local laws and standards.
   - Addressing conflicts arising from cross-border data flows.
2. **Bias Mitigation**:
   - Identifying and mitigating biases in datasets to ensure equitable outcomes in AI applications.
   - Developing tools to audit and correct biased algorithms.

3. **Ethical Data Sharing**:
    - Balancing the benefits of data sharing with privacy and security concerns.
    - Establishing standards for anonymization and consent-based data use.
4. **Data Valuation Models**:
    - Researching methods to quantify the value of data assets for businesses and individuals.
    - Exploring compensation models for users contributing data to platforms.

4. Sustainability and Green ICT

The environmental impact of ICT infrastructure has become a critical area of concern. Research in sustainability focuses on reducing energy consumption, minimizing electronic waste, and developing eco-friendly technologies.

Key Research Areas in Sustainability:

1. **Energy-Efficient Computing**:
    - Designing algorithms and hardware that consume less power.
    - Exploring renewable energy sources for data centers and ICT infrastructure.
2. **E-Waste Management**:
    - Developing recycling methods to recover valuable materials from discarded electronics.
    - Promoting circular economy principles in device manufacturing.
3. **Green Data Centers**:
    - Optimizing cooling systems and energy use in data centers.
    - Implementing AI to monitor and manage energy consumption.
4. **Carbon Footprint Reduction**:
    - Assessing the environmental impact of emerging technologies like blockchain.
    - Innovating low-impact solutions for ICT deployment in rural and remote areas.

5. Digital Inclusion and Accessibility

Ensuring equitable access to ICT resources remains a global challenge. Research in digital inclusion aims to bridge the digital divide and make technology accessible to diverse populations.

Key Research Areas in Digital Inclusion:

1. **Rural Connectivity**:
    - Exploring cost-effective solutions for expanding broadband access in underserved areas.
    - Research into satellite-based internet services like Starlink.

2. **Accessible Technologies**:
   - Designing ICT tools for individuals with disabilities, such as screen readers, speech-to-text applications, and tactile devices.
   - Developing universal design principles for software and hardware.
3. **Affordable Devices and Services**:
   - Researching low-cost computing devices for educational and vocational use.
   - Creating subsidized internet plans to enhance affordability.
4. **Digital Literacy**:
   - Evaluating the effectiveness of digital literacy programs.
   - Creating scalable training modules tailored to different demographics.

6. Emerging Technologies and Applications

The rapid emergence of technologies like blockchain, 5G, and quantum computing presents unique research challenges and opportunities.

Key Research Areas in Emerging Technologies:

1. **Blockchain Technology**:
   - Exploring its use in supply chain transparency, secure voting systems, and decentralized finance (DeFi).
   - Addressing scalability issues and energy consumption concerns.
2. **5G and Beyond**:
   - Researching applications of 5G in smart cities, telemedicine, and autonomous vehicles.
   - Addressing security concerns in ultra-reliable, low-latency communications.
3. **Quantum Computing**:
   - Investigating practical applications for quantum algorithms in cryptography, logistics, and artificial intelligence.
   - Developing error correction techniques to improve quantum systems' reliability.
4. **Edge and Fog Computing**:
   - Enhancing the efficiency of data processing near its source.
   - Exploring its applications in IoT ecosystems and real-time analytics.

7. Human-Technology Interaction

The increasing integration of technology into daily life necessitates research into its impact on human behavior and well-being.

Key Research Areas in Human-Technology Interaction:

1. **Digital Well-Being:**
    - Understanding the psychological effects of technology use, including screen time and social media addiction.
    - Designing tools to promote healthier tech habits.
2. **User Experience (UX):**
    - Researching intuitive design principles to enhance usability.
    - Incorporating feedback loops to continuously improve ICT interfaces.
3. **Workplace Transformation:**
    - Analyzing the impact of remote work and virtual collaboration tools on productivity and mental health.
    - Exploring hybrid work models for better employee satisfaction.

Chapter 4

Transition of Traditional Media Transition to New Media

Introduction

The evolution of communication and information-sharing practices over the last few decades has been monumental, largely due to the transformative influence of new media. What began as traditional forms of media—newspapers, radio, television, and print—has now transitioned into a dynamic, interactive, and user-centered ecosystem that encompasses the internet, social media platforms, blogs, streaming services, and digital multimedia. This shift has not only altered how information is produced and consumed but has also reshaped the very fabric of social, cultural, and economic interactions on a global scale.

The advent of new media represents more than just technological advancement; it is a paradigm shift in the way societies engage with knowledge, information, and communication tools. Traditional media, characterized by one-way communication and centralized control, operated within well-defined boundaries of production and dissemination. These mediums were predominantly linear, with content creators having near-total control over what audiences saw, heard, or read. For example, newspapers were curated by editors, and television programs were scheduled by broadcasters, leaving audiences as passive recipients of information.

However, the digital revolution ushered in by the internet has disrupted these traditional models. The transition from traditional to new media is not merely a replacement but a redefinition of media's role in society. New media fosters interactivity, user-generated content, and real-time engagement, blurring the lines between producers and consumers of content. The audience is no longer a passive entity; they are active participants who create, share, and comment on content, shaping narratives in unprecedented ways. Blogs, podcasts, YouTube channels, and social networking sites like Facebook, Twitter, and Instagram exemplify how individuals now wield significant power to influence public discourse, challenge traditional hierarchies, and create decentralized communication networks.

The phenomenon of knowledge management and archiving within the new media landscape further underscores its transformative impact. The digitization of information has enabled vast amounts of data to be stored, categorized, and retrieved with incredible efficiency. Unlike traditional methods that relied on physical records and limited accessibility, digital technologies offer seamless access to archives, allowing individuals and organizations to manage knowledge more effectively. Platforms like Google Drive, OneDrive, and open-access repositories have democratized access to information, fostering a culture of knowledge sharing. Moreover, these digital systems facilitate collaboration on a global scale, enabling researchers, educators, and businesses to pool resources and innovate across boundaries.

DOI: 10.4324/9781003750444-4

Networks form the backbone of new media, facilitating its expansive reach and interconnectivity. The internet, as the most significant network, has become the central hub for new media activities. It connects billions of devices, enabling instant communication, file sharing, and content dissemination. The transition to high-speed broadband and the proliferation of mobile internet have further accelerated this connectivity. Networks do not merely serve as conduits for information transfer; they represent complex systems that define how individuals and organizations interact in the digital age. Social networks, in particular, have redefined relationships, forming virtual communities that transcend geographic and cultural barriers. Platforms like LinkedIn for professional networking, TikTok for creative expression, and Reddit for interest-based discussions illustrate the diversity and adaptability of these networks.

Social media, a quintessential component of new media, exemplifies both its advantages and limitations. On the positive side, social media has democratized access to information, given a voice to marginalized communities, and created opportunities for social activism. Movements such as #BlackLivesMatter, #MeToo, and climate change advocacy campaigns demonstrate how social media can mobilize collective action and amplify important issues. It allows users to connect with like-minded individuals, fostering a sense of community and shared purpose. For businesses and marketers, social media offers unparalleled tools for audience engagement, targeted advertising, and real-time feedback, transforming how brands interact with consumers. Yet, the limitations of social media are equally significant. The rise of misinformation, echo chambers, and algorithm-driven content raises concerns about its impact on public discourse and mental health. Platforms prioritize engagement metrics, often at the expense of accuracy and balanced viewpoints, leading to the proliferation of fake news and polarizing content. The addictive nature of social media, coupled with its ability to manipulate attention, has sparked debates about ethical technology design and the long-term psychological effects of constant connectivity. Privacy concerns are another critical issue, with users often unaware of the extent to which their data is collected, analyzed, and monetized.

The transition from traditional to new media has also challenged the concepts of authenticity, credibility, and authority. While traditional media was often perceived as a gatekeeper of verified information, new media platforms lack the same level of editorial oversight. This has led to a democratization of voices, but also a rise in unreliable sources. In the era of citizen journalism, anyone with a smartphone can report events, bypassing traditional news channels. While this has empowered individuals and exposed injustices, it has also blurred the lines between fact and opinion, creating a landscape where discerning truth requires critical thinking and digital literacy. New media has also revolutionized cultural production and consumption. The shift from physical to digital formats has given rise to streaming services like Netflix, Spotify, and YouTube, where users can access vast libraries of content on demand. This has disrupted traditional industries such as television, film, and music, which once relied on physical distribution and fixed schedules. Furthermore, digital platforms have enabled the emergence of independent creators, from vloggers and podcasters to indie game developers and digital artists, who can reach global audiences without the need for intermediaries. This has created a more diverse and inclusive media ecosystem, where voices that were previously marginalized can now thrive.

However, this digital transformation also comes with significant challenges. The decline of traditional media businesses, such as newspapers and television broadcasters, has raised concerns about the sustainability of investigative journalism and quality content production. The reliance on digital advertising, dominated by tech giants like Google and Facebook, has squeezed revenue streams for smaller media organizations. Additionally, the algorithms that drive content discovery often prioritize sensationalism over substance, leading to a culture of clickbait and reduced attention spans. The integration of new media into everyday life has also led to the phenomenon of "always-on" connectivity, where individuals are perpetually engaged with digital devices. This has profound implications for work-life balance, mental health, and interpersonal relationships. While new media facilitates remote work, online learning, and telemedicine, it also blurs the boundaries between personal and professional spaces, creating pressures to remain constantly productive and available. The cultural shift towards digital dependence has prompted discussions about digital detoxes, mindfulness, and the need to establish healthy technology habits.

From an economic perspective, new media has reshaped markets, creating opportunities for innovation and entrepreneurship. The rise of e-commerce platforms like Amazon, eBay, and Shopify has transformed retail, enabling small businesses to reach global markets. Social media influencers and content creators have established new revenue models through sponsorships, ad revenues, and merchandise sales. However, these shifts also raise questions about labor rights, job security, and the gig economy. As traditional employment structures give way to freelance and contract work in the digital age, policymakers and industry leaders must address the implications for workers' rights and benefits. The global nature of new media has also amplified cross-cultural interactions, fostering greater understanding and collaboration but also exposing tensions and misunderstandings. Digital platforms serve as arenas for cultural exchange, where users can share art, music, and traditions with global audiences. Yet, they can also become battlegrounds for cultural appropriation, misinformation, and cyber conflicts. The challenge lies in leveraging new media to promote inclusivity and dialogue while mitigating the risks of polarization and cultural erosion.

In education, new media has revolutionized how knowledge is delivered and consumed. Online learning platforms like Coursera, Khan Academy, and edX offer courses from leading institutions, making education accessible to millions worldwide. Digital tools such as virtual classrooms, e-books, and interactive simulations enhance learning experiences, particularly in remote or underserved areas. At the same time, the digital divide persists, with disparities in access to technology and connectivity preventing equitable participation in new media-driven opportunities. The rise of new media has implications for governance and civic engagement as well. Governments increasingly use digital platforms to communicate with citizens, deliver services, and engage in public diplomacy. E-governance initiatives, such as online tax filing, digital voting, and public grievance portals, streamline administrative processes and enhance transparency. However, the misuse of new media for propaganda, surveillance, and cyber warfare highlights the need for robust regulations and international cooperation to ensure the ethical use of technology.

A. Traditional Media Transition to New Media

The transition from traditional media to new media represents a transformative period in the history of communication, reshaping how individuals, organizations, and societies interact with information. This shift has been driven by technological advancements, changing consumer behavior, and the ubiquity of the internet and digital devices. Traditional media—characterized by print, radio, television, and film—was once the primary mode of information dissemination, defined by one-way communication and centralized control. In contrast, new media offers an interactive, decentralized, and dynamic environment that empowers users as both consumers and creators of content. The journey from traditional to new media is not merely a technological evolution but a cultural and structural shift that has redefined how information is created, distributed, and consumed.

The Foundations of Traditional Media

Traditional media, also known as "old media," encompasses communication channels such as newspapers, magazines, radio, and television that dominated the information and entertainment landscape throughout the 20th century. These mediums formed the backbone of mass communication, serving as the primary sources of news, cultural narratives, and public discourse for decades. Rooted in physical infrastructure and centralized production, traditional media was characterized by its reliance on one-way communication and its role as a gatekeeper of information. Despite its limitations, traditional media significantly influenced societal structures, public opinion, and cultural development.

Defining Characteristics of Traditional Media

The traditional media landscape was shaped by unique attributes that defined its operational mechanisms, audience interaction, and cultural influence. These attributes, while effective in their time, also highlighted the limitations of old media in adapting to the dynamic demands of a rapidly digitizing world.

1. Linear Communication

- **Unidirectional Flow of Information**: Traditional media operated on a sender-receiver communication model. In this model, information flowed in one direction—from producers, such as journalists, broadcasters, and editors, to audiences. The content was created, curated, and disseminated by a select group of professionals, leaving audiences with little to no opportunity to provide feedback or influence the narrative.
- **Audience Passivity**: Audiences consumed content passively, as there were few mechanisms for interaction or engagement. For instance:
    - Newspapers and magazines provided written articles without immediate avenues for reader response.
    - Radio and television broadcasts followed fixed schedules, with audiences having no control over the timing or content of programs.
- **Impact**: This unidirectional communication model reinforced the authority of traditional media as the primary gatekeeper of information, shaping public opinion without significant input from the audience.

2. Gatekeeping

- **Centralized Control of Content**: Media organizations acted as gatekeepers, determining what information was worthy of publication or broadcast. This process involved rigorous editorial oversight, ensuring that content adhered to journalistic and ethical standards. Editors played a crucial role in shaping the narrative by selecting topics, framing stories, and prioritizing coverage.

- **Limited Perspectives**: While gatekeeping ensured credibility and quality, it also restricted the diversity of perspectives. Content often reflected the priorities and biases of the media organization, leaving alternative voices and marginalized communities underrepresented.
- **Example**:
  - National newspapers like *The New York Times* or *The Times of India* set the agenda for public discourse by focusing on specific issues, while sidelining others deemed less newsworthy.
- **Implications**: Gatekeeping cemented the role of traditional media as a trusted source of information but also highlighted its limitations in fostering a pluralistic media environment.

3. Mass Audiences

- **Broad Appeal**: Traditional media aimed to cater to a wide audience demographic. Content was designed to appeal to the largest possible number of viewers, readers, or listeners. For example:
  - Television programs were crafted to attract family audiences during prime-time slots.
  - Newspapers provided a mix of news, sports, entertainment, and opinion columns to cover varied interests.
- **Generalized Content**: The focus on mass appeal often resulted in content that was broad and inclusive but lacked the personalization and niche targeting that new media platforms later offered.
- **Cultural Impact**:
  - Traditional media played a significant role in fostering cultural cohesion by presenting shared narratives and collective experiences. For instance, live broadcasts of significant events, such as moon landings or national elections, united audiences across geographic and social divides.
- **Limitations**: The one-size-fits-all approach meant that traditional media struggled to cater to specific audience preferences or niche interests.

4. Time and Space Constraints

- **Physical Distribution**: Print media relied on physical infrastructure for production and distribution, which limited its geographic reach. Newspapers and magazines were bound by delivery networks, and their availability was often restricted to urban and semi-urban areas, leaving rural communities underserved.
- **Broadcast Schedules**: Radio and television adhered to fixed programming schedules. Audiences had to tune in at specific times to access content, such as news broadcasts, entertainment shows, or sports events. Missing a broadcast often meant missing the content entirely, as on-demand access was unavailable.
- **Example**:
  - Prime-time television slots dictated when families could gather to watch popular shows or important news events.

    - Morning newspaper delivery dictated when individuals accessed the latest news.

- **Challenges:** These constraints on time and space created inefficiencies in content delivery and accessibility, paving the way for more flexible digital alternatives.

The Role of Traditional Media

Despite its inherent limitations, traditional media played a pivotal role in shaping society. Its influence extended across politics, culture, education, and social dynamics, contributing to its stature as a cornerstone of public life.

1. **Shaping Public Opinion**:
    - Traditional media was instrumental in informing the public about significant events, shaping perceptions, and driving political discourse. Investigative journalism, for instance, exposed corruption and injustices, holding power structures accountable.
2. **Fostering Cultural Identity**:
    - By showcasing national events, cultural heritage, and shared experiences, traditional media created a sense of unity and belonging among diverse populations.
3. **Educating the Masses**:
    - Public service announcements, documentaries, and educational programming helped raise awareness about critical issues, such as public health, environmental conservation, and literacy.
4. **Entertainment**:
    - Traditional media provided escapism through films, television dramas, and music, influencing cultural trends and individual lifestyles.

The Decline of Traditional Media

The dominance of traditional media began to erode with the advent of the internet and digital technologies in the late 20th century. This decline was marked by several key developments:

1. **Rise of New Media**:
    - Digital platforms offered interactive, on-demand, and user-centric experiences that traditional media could not match. Social media, streaming services, and blogs empowered audiences to access, create, and share content freely.
2. **Economic Challenges**:
    - Declining circulation and advertising revenues forced newspapers and magazines to transition to online formats. Television and radio faced competition from streaming services and podcasts.
3. **Changing Consumer Preferences**:
    - Audiences sought personalized and niche content, shifting away from the generalized offerings of traditional media.

4. **Technological Innovations**:
   - High-speed internet, mobile devices, and advanced algorithms revolutionized content delivery, rendering traditional distribution methods inefficient.

Legacy and Influence of Traditional Media

While traditional media's dominance has waned, its legacy continues to shape the media landscape. Established media organizations have adapted to the digital age by embracing hybrid models that combine traditional journalistic principles with the interactivity of new media. The credibility, professionalism, and ethical standards of traditional media remain benchmarks for content creation in the digital era.

Drivers of the Transition to New Media

The transition from traditional to new media marks one of the most profound transformations in the history of communication and information dissemination. This shift has been fueled by a convergence of technological, social, and economic forces that have fundamentally reshaped how people create, access, and engage with content. Each of these drivers has played a pivotal role in enabling the dynamic, interactive, and participatory ecosystem that defines new media. Unlike traditional media's linear, one-way communication model, new media thrives on decentralization, interactivity, and user empowerment.

1. Technological Advancements

The foundation of the transition to new media lies in rapid technological advancements that have revolutionized how information is created, shared, and consumed.

- **The Internet Revolution**:
  - The development of the internet has been the single most transformative factor in the shift to new media. By providing a global, decentralized network for communication, the internet has eliminated many of the barriers associated with traditional media.
  - Early internet protocols, such as email and Usenet, laid the groundwork for modern digital communication, while the advent of the World Wide Web in the 1990s introduced user-friendly platforms for accessing and sharing content.
- **Mobile Devices and Ubiquitous Connectivity**:
  - The proliferation of mobile devices, including smartphones and tablets, has brought new media into the hands of billions of users worldwide. Mobile technology enables users to access content on the go, fostering a culture of always-on connectivity.
  - High-speed broadband and 4G/5G networks have further enhanced accessibility, allowing seamless streaming, video calls, and real-time interactions.
- **Digital Content Creation Tools**:
  - Advances in software and hardware have democratized content creation. Tools

such as video editing software, graphic design applications, and user-friendly website builders enable individuals with limited technical expertise to produce professional-quality content.

  - Platforms like YouTube, TikTok, and Instagram have harnessed these tools, empowering users to become creators and influencers.

- **Emergence of Social Media Platforms**:
  - Social media platforms like Facebook, Twitter, and LinkedIn have redefined how people connect, share information, and engage with content. These platforms prioritize user-generated content and encourage participatory communication.

2. Changing Consumer Behavior

The expectations and behaviors of media consumers have evolved significantly, driven by the possibilities offered by new media technologies.

- **Demand for Interactivity**:
  - Unlike traditional media, which offered a passive consumption experience, new media emphasizes interactivity. Users now expect to comment on articles, share videos, and participate in online discussions.
  - Live streaming, real-time polling, and interactive webinars exemplify how new media engages users in active communication.
- **Shift to On-Demand Content**:
  - Audiences increasingly favor on-demand access to content, a stark contrast to the fixed schedules of traditional media. Streaming platforms like Netflix, Spotify, and YouTube cater to this preference, allowing users to consume content at their convenience.
  - Podcasts, audiobooks, and digital libraries have also gained popularity for their flexibility and accessibility.
- **Personalization and Control**:
  - Consumers value platforms that provide tailored experiences. Algorithms on platforms like Netflix, Amazon, and Facebook analyze user behavior to recommend personalized content, enhancing user satisfaction and engagement.
  - The ability to control what, when, and how content is consumed has empowered users and challenged the one-size-fits-all approach of traditional media.
- **The Role of Younger Generations**:
  - Younger demographics, often referred to as "digital natives," have grown up with the internet and mobile technology. They prefer platforms that align with their tech-savvy lifestyles, such as Snapchat, TikTok, and Discord.
  - These generations are also more likely to engage with multimedia content and participate in online communities.

3. Cost and Accessibility

The cost-effectiveness and accessibility of new media have significantly contributed to its widespread adoption.

- **Lower Barriers to Entry**:
    - In traditional media, creating and distributing content required significant resources, such as printing presses, broadcast equipment, and distribution networks. New media has drastically reduced these costs, enabling individuals and small organizations to compete with established media companies.
    - For example, a blogger with minimal investment can reach a global audience, while independent filmmakers can distribute their work through streaming platforms.
- **Affordable Technology**:
    - The affordability of smartphones, tablets, and laptops has made digital technology accessible to a broader demographic, including low-income and rural populations.
    - Initiatives to provide free or subsidized internet access, such as public Wi-Fi hotspots and government-led connectivity programs, have further expanded the reach of new media.
- **Crowdsourcing and Crowdfunding**:
    - New media platforms have popularized funding models like crowdsourcing and crowdfunding, enabling creators to finance their projects directly through audience contributions.
    - Platforms like Patreon, Kickstarter, and GoFundMe exemplify how creators can sustain their work without relying on traditional advertisers or publishers.

4. Globalization

New media has transcended geographic and cultural boundaries, fostering a global exchange of ideas, information, and innovation.

- **Cross-Border Connectivity**:
    - The interconnected nature of the internet allows users from different parts of the world to communicate and collaborate in real time. This has facilitated the growth of global online communities, ranging from professional networks to fan-based groups.
    - Platforms like Reddit and GitHub demonstrate how new media supports knowledge sharing and collective problem-solving on a global scale.
- **Emergence of a Global Digital Culture**:
    - New media has contributed to the development of a shared global culture characterized by memes, viral content, and digital trends. While this cultural exchange promotes inclusivity and understanding, it also raises questions about cultural homogenization and the loss of local traditions.

- For instance, global streaming platforms like Spotify and Netflix often dominate local markets, influencing regional media consumption patterns.

- **Opportunities for Cross-Border Collaboration**:
  - Artists, entrepreneurs, and academics can leverage new media to collaborate across borders, creating innovative projects and businesses that were previously impossible.
  - Example: Online learning platforms like Coursera and edX connect learners and educators worldwide.

5. Economic Pressures on Traditional Media

The rise of new media has disrupted traditional media's economic models, compelling legacy organizations to adapt or face obsolescence.

- **Declining Revenues**:
  - Advertising, a key revenue stream for traditional media, has shifted to digital platforms. Companies like Google, Facebook, and Amazon dominate the digital advertising market, leaving traditional media with a smaller share.
  - Print circulation for newspapers and magazines has steadily declined, further exacerbating financial challenges.
- **Rise of Digital Advertising**:
  - Digital platforms offer advertisers precise targeting capabilities based on user data, making them more appealing than traditional media. Algorithms and analytics enable brands to measure the effectiveness of their campaigns in real time.
- **Shift to Subscription Models**:
  - To counter declining ad revenues, many traditional media outlets have adopted subscription-based models. Digital paywalls and memberships are now common in the news industry, with outlets like *The New York Times* and *The Guardian* relying on reader contributions.
- **Challenges of Adapting to Digital**:
  - Legacy media organizations often face difficulties in transitioning to digital formats. These include integrating new technologies, retraining staff, and managing the cultural shift required to thrive in a digital-first environment.

Characteristics of New Media

New media is a transformative evolution in communication that stands apart from traditional media by its dynamic, participatory, and decentralized nature. It has redefined how people create, share, and consume information, fostering greater engagement, diversity, and interactivity. The emergence of digital platforms, social media, and streaming services has enabled a more user-centric approach to media, challenging the hierarchical structure of traditional gatekeepers and opening new avenues for expression and collaboration. The defining characteristics of new media underscore its innovative role in reshaping the communication landscape. These attributes—interactivity, user-generated content, on-demand access, personalization, decentralization, and convergence—collectively highlight its disruptive potential and its profound impact on culture, society, and industry.

1. Interactivity

One of the most defining features of new media is its inherent interactivity, which contrasts sharply with the passive, one-way communication model of traditional media. Interactivity enables users to engage actively with content, fostering dynamic exchanges between creators and audiences.

- **Two-Way and Multi-Directional Communication**:
    - New media platforms facilitate conversations and interactions between users and creators, breaking the barriers of linear communication.
    - Users can comment on articles, share videos, react to posts, and participate in live discussions, creating a continuous feedback loop.
- **Engagement and Community Building**:
    - Interactivity encourages the formation of online communities where users with shared interests can collaborate and exchange ideas.
    - For example, platforms like Reddit and Discord serve as hubs for niche communities engaging in active discussions.
- **Examples**:
    - YouTube allows viewers to comment, like, and share videos, enabling creators to respond directly to their audiences.
    - Instagram Stories invite real-time interaction through polls, Q&A features, and direct messages.
- **Impact**:
    - Interactivity enhances user engagement, fostering loyalty and participation.
    - It democratizes communication by giving audiences a voice and a role in shaping narratives.

2. User-Generated Content

New media has empowered individuals to become creators, breaking the monopoly of traditional media over content production and distribution. This democratization of media production has transformed the roles of consumers and producers, giving rise to the concept of "prosumers" (producers and consumers).

- **Creation Without Gatekeepers**:
    - Platforms like YouTube, Instagram, and TikTok enable users to produce and share their own content without the need for approval from traditional gatekeepers such as editors or broadcasters.
    - This has opened opportunities for diverse voices and perspectives to thrive.

- **Citizen Journalism and Influencer Culture**:
  - The rise of user-generated content has led to the emergence of citizen journalists who report on events in real time, often bypassing traditional news outlets.
  - Influencer marketing has become a powerful industry, with creators leveraging their personal brands to connect with audiences and promote products.
- **Independent Creators and Entrepreneurs**:
  - Artists, musicians, writers, and filmmakers can now reach global audiences directly through platforms like Patreon, Substack, and Vimeo.
  - These platforms enable creators to monetize their work through subscriptions, donations, and merchandise sales.
- **Impact**:
  - User-generated content fosters diversity in media representation and challenges traditional hierarchies.
  - It has also disrupted industries by creating alternative revenue models and redefining audience expectations.

3. On-Demand Access

New media offers unparalleled flexibility in content consumption, enabling users to access information and entertainment at their convenience. This on-demand nature represents a significant departure from the fixed schedules of traditional media.

- **Breaking Time and Space Constraints**:
  - Users no longer need to adhere to broadcast schedules or physical distribution methods.
  - Content is available anytime and anywhere, as long as there is an internet connection.
- **Streaming and Subscription Models**:
  - Platforms like Netflix, Hulu, and Spotify have revolutionized how media is consumed by offering vast libraries of content accessible on demand.
  - Podcasts and e-books have similarly transformed the consumption of audio and written content.
- **Examples**:
  - A user can binge-watch an entire series on Netflix instead of waiting for weekly episodes.
  - Spotify allows users to create personalized playlists and access millions of songs instantly.
- **Impact**:
  - On-demand access caters to individual preferences and lifestyles, enhancing user satisfaction.

   - It has also disrupted traditional media industries, forcing them to adopt similar models to remain competitive.

4. Personalization

Personalization is a hallmark of new media, enabled by advanced algorithms and data analytics that tailor content to individual preferences. This customization enhances user engagement and creates a sense of relevance and connection.

- **Algorithm-Driven Recommendations**:
   - Platforms analyze user behavior, such as browsing history, likes, and search queries, to recommend content that aligns with their interests.
   - Examples include personalized news feeds on Facebook and curated playlists on Spotify.
- **Targeted Advertising**:
   - Personalization extends to advertising, where brands can deliver highly targeted messages to specific demographics based on user data.
   - This approach increases the effectiveness of marketing campaigns.
- **User-Centric Experiences**:
   - Personalization allows users to control their media consumption, choosing what to watch, read, or listen to based on their unique tastes.
- **Impact**:
   - While personalization improves user experience, it also raises concerns about privacy, data security, and the creation of "echo chambers" that limit exposure to diverse viewpoints.

5. Decentralization

Unlike traditional media, which is controlled by centralized entities such as publishing houses or broadcasting corporations, new media operates in a decentralized framework. This decentralization has profound implications for accessibility, diversity, and power dynamics.

- **Distributed Networks**:
   - Content in new media is distributed across a network of users, platforms, and servers, reducing the control of a single entity.
   - Peer-to-peer systems, such as file-sharing networks, exemplify this decentralized structure.
- **Empowerment of Marginalized Voices**:
   - Decentralization provides a platform for underrepresented communities to share their stories and perspectives without gatekeeping.
   - Social media movements like #MeToo and #BlackLivesMatter illustrate how decentralized platforms can amplify marginalized voices.

- **Challenges to Authority**:
  - The decentralization of new media has disrupted traditional power structures, challenging established media organizations, governments, and corporations.
- **Impact**:
  - Decentralization fosters a more inclusive and participatory media environment but also poses challenges in regulating content and ensuring accountability.

6. Convergence

New media is characterized by the convergence of multiple forms of communication, integrating text, audio, video, graphics, and interactivity into a single platform. This fusion enhances the versatility and appeal of new media.

- **Blurring Media Boundaries**:
  - Social media platforms like Twitter and Instagram combine elements of blogging, photography, and broadcasting, creating a multimedia experience.
  - Video game streaming platforms like Twitch integrate live video, chat, and community engagement.
- **Cross-Media Integration**:
  - New media platforms enable seamless integration with other technologies, such as augmented reality (AR), virtual reality (VR), and the Internet of Things (IoT).
  - Example: Snapchat filters use AR technology to create interactive experiences.
- **Enhanced Storytelling**:
  - Convergence allows for innovative storytelling methods, combining visuals, sound, and interactivity to engage audiences more deeply.
  - Examples include interactive documentaries and multimedia journalism.
- **Impact**:
  - Convergence has expanded the creative possibilities of media, enabling more immersive and engaging experiences.
  - It has also blurred the lines between different media industries, fostering collaboration and innovation.

Impact of the Transition from Traditional to New Media

The transition from traditional to new media represents a seismic shift in how information is created, shared, and consumed. This change has had profound implications for industries, societies, and individual behaviors, reshaping the global communication ecosystem. Traditional media, once characterized by centralized control and passive audiences, has given way to a dynamic, participatory environment enabled by digital platforms, algorithms, and mobile connectivity. While new media has brought about significant advancements and opportunities, it has also introduced challenges that require careful consideration.

1. Media Industry Disruption

The rise of new media has disrupted traditional media industries, forcing legacy organizations to adapt or face obsolescence. This disruption has transformed business models, content delivery, and audience engagement.

- **Adoption of Digital Strategies**:
  - Traditional media organizations have embraced digital-first approaches to stay relevant. This includes creating online editions of newspapers, implementing paywalls, and offering subscription-based services.
  - Examples include *The New York Times* and *The Washington Post*, which have successfully transitioned to digital platforms, generating substantial revenue from online subscribers.
- **Rise of Streaming Services**:
  - Platforms like Netflix, Amazon Prime Video, and Hulu have revolutionized the television and film industries, shifting audiences from scheduled programming to on-demand content.
  - The decline of cable television subscriptions, often referred to as "cord-cutting," reflects this trend.
- **Transformation of Print Media**:
  - Print media has faced significant challenges due to declining circulation and advertising revenues. Digital publishing has emerged as a viable alternative, with platforms like Medium and Substack offering opportunities for independent writers.
- **Implications**:
  - Media organizations must continuously innovate to retain audience attention in an increasingly competitive digital landscape.
  - The democratization of content creation has also challenged traditional media's role as the primary gatekeeper of information.

2. Empowerment of Individuals

One of the most transformative aspects of new media is its ability to empower individuals by providing them with tools to create, share, and access content.

- **Democratization of Information**:
  - New media platforms have removed traditional barriers to entry, allowing individuals to share their stories, ideas, and talents without relying on established media institutions.
  - Examples include citizen journalism, where ordinary people report news through platforms like Twitter and YouTube.

- **Amplification of Marginalized Voices**:
    - Social media has provided a platform for underrepresented communities to raise awareness about social justice issues, promote cultural identities, and challenge systemic inequalities.
    - Movements such as #MeToo, #BlackLivesMatter, and climate activism campaigns demonstrate the power of new media to drive societal change.
- **Personal Branding and Influence**:
    - Platforms like Instagram, TikTok, and LinkedIn enable individuals to build personal brands and influence public opinion. Influencers and content creators have emerged as powerful voices, shaping trends and consumer behavior.
- **Implications**:
    - While empowerment fosters creativity and inclusivity, it also places greater responsibility on individuals to critically evaluate and disseminate information.

3. Cultural Shifts

The transition to new media has redefined cultural dynamics, blending local traditions with global influences and creating new forms of expression.

- **Blending of Cultures**:
    - Digital platforms facilitate the exchange of ideas, art, and traditions across borders. For example, music streaming services expose users to diverse genres, while platforms like TikTok popularize dance trends and cultural practices globally.
- **Emergence of Digital Subcultures**:
    - New media has given rise to niche communities and subcultures, from gaming enthusiasts on Twitch to fanfiction writers on platforms like Wattpad.
- **Concerns About Cultural Homogenization**:
    - While globalization fosters inclusivity, it also raises concerns about the erosion of local cultures and traditions. Dominance by Western digital platforms often leads to the prioritization of global trends over regional identities.
- **Implications**:
    - Balancing global connectivity with the preservation of cultural diversity requires intentional efforts from content creators, platforms, and policymakers.

4. Economic Implications

The economic impact of new media extends across industries, fostering innovation while raising concerns about monopolistic practices and income inequality.

- **Opportunities in the Gig Economy**:
    - Platforms like YouTube, Twitch, and Patreon have created new revenue streams for independent creators, enabling them to monetize their skills and content.

  - Freelance platforms like Fiverr and Upwork allow professionals to offer services directly to global clients.
- **Entrepreneurial Innovation**:
  - New media has lowered the barriers for launching businesses, with e-commerce platforms like Shopify and Etsy enabling small-scale entrepreneurs to reach global audiences.
- **Dominance of Tech Giants**:
  - Companies like Google, Facebook, and Amazon dominate the digital economy, leveraging their market power to control advertising revenues, data collection, and content distribution.
  - Concerns about anti-competitive practices and data monopolies have led to calls for greater regulation.
- **Implications**:
  - While new media fosters economic opportunities, addressing issues of income disparity and monopolistic control is essential for equitable growth.

5. Challenges of Misinformation

The rise of new media has created an environment where misinformation, fake news, and conspiracy theories can spread rapidly, undermining trust in information sources.

- **Lack of Editorial Oversight**:
  - Unlike traditional media, which employs rigorous editorial processes, new media platforms often lack mechanisms to verify the accuracy of user-generated content.
  - Viral misinformation campaigns have influenced public opinion on issues ranging from elections to public health.
- **Algorithmic Amplification**:
  - Social media algorithms prioritize content that drives engagement, often amplifying sensational or polarizing material, regardless of its accuracy.
- **Efforts to Combat Misinformation**:
  - Platforms like Facebook and Twitter have implemented fact-checking initiatives and content moderation policies.
  - Digital literacy campaigns aim to educate users about identifying reliable sources and critical evaluation of information.
- **Implications**:
  - Combating misinformation requires a multi-faceted approach involving platforms, governments, and civil society.

6. Privacy and Security Concerns

The reliance of new media on user data for personalization and advertising has raised significant ethical and security challenges.

- **Data Privacy Issues:**
  - Platforms collect vast amounts of personal data to target users with tailored content and advertisements. This practice raises concerns about consent and transparency.
  - High-profile data breaches and scandals, such as the Cambridge Analytica incident, have highlighted the risks of inadequate data protection.
- **Surveillance and Tracking:**
  - Governments and corporations increasingly use digital tools for surveillance, raising concerns about individual freedoms and civil liberties.
  - Facial recognition, location tracking, and social media monitoring are examples of practices that challenge privacy norms.
- **Balancing Innovation and Regulation:**
  - Policymakers face the challenge of fostering technological innovation while protecting user rights.
  - Laws such as the General Data Protection Regulation (GDPR) in the European Union aim to create frameworks for responsible data use.
- **Implications:**
  - Ensuring privacy and security in the digital age requires robust regulatory measures, user education, and ethical practices by technology companies.

Chapter 5

Knowledge Management and Archiving

Knowledge management and archiving represent critical components of the modern information ecosystem, especially in the era of new media. As digital technologies proliferate, the ability to efficiently create, store, organize, retrieve, and share information has become essential for organizations, individuals, and societies at large. Knowledge management focuses on leveraging organizational knowledge to enhance decision-making, innovation, and productivity, while archiving ensures the long-term preservation and accessibility of information for historical, cultural, and operational purposes.

The Role of Archiving in Knowledge Management

Archiving complements knowledge management by ensuring the long-term preservation and accessibility of information. Archives serve as repositories for historical records, legal documents, cultural artifacts, and organizational knowledge.

Knowledge Management

Knowledge Management (KM) is a strategic process that focuses on capturing, organizing, sharing, and analyzing knowledge within an organization or system to achieve specific goals. It aims to leverage the collective intelligence of an organization to drive innovation, improve decision-making, enhance operational efficiency, and preserve institutional knowledge. KM integrates both explicit knowledge, which is easily documented and shared, and tacit knowledge, which is deeply rooted in personal experiences and insights and often challenging to formalize. In an era characterized by rapid technological advancements and data-driven decision-making, KM plays a critical role in ensuring that organizations remain competitive, adaptable, and capable of leveraging their intellectual assets effectively.

Key Objectives of Knowledge Management

Knowledge Management serves as a cornerstone for organizational success, addressing various needs that range from enhancing decision-making processes to preserving institutional memory. Each objective underscores the importance of systematic knowledge handling in achieving organizational efficiency and growth.

1. Enhancing Decision-Making

One of the most significant objectives of KM is to improve the quality of decision-making by

DOI: 10.4324/9781003750444-5

ensuring that relevant, accurate, and up-to-date information is readily accessible to decision-makers.

- **How It Works:**
  - KM systems integrate information from multiple sources, including internal databases, market research, and customer feedback, to provide a comprehensive knowledge base.
  - By offering easy access to historical data, industry insights, and predictive analytics, KM enables informed and timely decisions.
- **Example:**
  - A healthcare organization might use a KM system to store and share patient care protocols, clinical guidelines, and research findings among medical staff. This ensures that doctors and nurses can make evidence-based decisions quickly, improving patient outcomes.
- **Impact:**
  - Reduces errors in decision-making by minimizing reliance on incomplete or outdated information.
  - Enhances the ability to respond proactively to market changes, customer needs, and operational challenges.

2. Promoting Innovation

KM fosters a culture of collaboration and continuous learning, enabling teams to build on existing knowledge and generate creative solutions. By breaking down silos and encouraging cross-functional interaction, KM creates an environment conducive to innovation.

- **How It Works:**
  - KM platforms facilitate the sharing of ideas and best practices across departments, regions, and even organizations.
  - Collaborative tools such as shared workspaces, idea repositories, and brainstorming sessions help harness collective intelligence.
- **Example:**
  - Leading tech companies like Google and Microsoft leverage KM systems to drive research and development efforts. Employees use shared platforms to exchange ideas, refine prototypes, and collaborate on complex projects.
- **Impact:**
  - Accelerates the development of new products, services, and processes.
  - Encourages a mindset of experimentation and adaptability, essential for staying ahead in competitive markets.

3. Improving Efficiency

KM eliminates redundancies and inefficiencies by ensuring that knowledge is systematically stored, organized, and made accessible to the right people at the right time. This streamlines workflows and reduces time spent searching for information.

- **How It Works**:
  - Centralized knowledge repositories allow employees to quickly retrieve case studies, templates, and best practices.
  - Automated tagging and categorization ensure that information is easy to find and use.
- **Example**:
  - Consulting firms maintain robust KM systems to provide consultants with quick access to client case studies, industry reports, and project methodologies. This reduces the time required to prepare for new assignments.
- **Impact**:
  - Shortens project timelines by eliminating the need to "reinvent the wheel."
  - Increases productivity by enabling employees to focus on value-added activities rather than information retrieval.

4. Preserving Institutional Memory

Institutional memory refers to the collective knowledge and experiences accumulated by an organization over time. KM ensures that this invaluable resource is retained, even as employees leave or retire.

- **How It Works**:
  - KM systems capture tacit knowledge through mechanisms such as mentorship programs, interviews, and documentation of key processes.
  - Digital archives store historical data, policies, and project outcomes, making them accessible to future employees.
- **Example**:
  - A government agency might use KM tools to document policy decisions, implementation challenges, and lessons learned. These records help new employees understand the rationale behind past actions and avoid repeating mistakes.
- **Impact**:
  - Reduces the loss of critical knowledge due to employee turnover.
  - Enhances continuity and stability by preserving organizational wisdom and expertise.

Core Components of Knowledge Management

Effective KM involves integrating several core components that work together to capture, organize, and disseminate knowledge:

1. **Knowledge Capture**:
   - Techniques such as interviews, surveys, and digital documentation help gather explicit and tacit knowledge.
2. **Knowledge Organization**:
   - Metadata, taxonomies, and tagging systems categorize and structure knowledge for easy access.
3. **Knowledge Sharing**:
   - Tools such as intranets, wikis, and collaboration platforms facilitate the exchange of knowledge across teams.
4. **Knowledge Application**:
   - Insights from KM systems are used to guide decision-making, optimize processes, and drive innovation.

Core Processes in Knowledge Management

Effective knowledge management (KM) requires a structured approach to systematically handle organizational knowledge. This involves capturing valuable insights, organizing them for efficient retrieval, sharing them across teams, and applying them to achieve strategic goals. The success of KM systems lies in their ability to transform individual expertise and scattered information into a cohesive knowledge ecosystem that drives productivity, innovation, and informed decision-making. The core processes in knowledge management—**Knowledge Capture**, **Knowledge Organization**, **Knowledge Sharing**, and **Knowledge Application**—are interconnected, forming a continuous cycle that sustains the knowledge lifecycle within an organization.

1. Knowledge Capture

Definition: Knowledge capture involves identifying and collecting both explicit and tacit knowledge within an organization. Explicit knowledge includes documented information such as reports, manuals, and presentations, while tacit knowledge consists of the insights, skills, and experiences held by individuals that are often harder to articulate and share.

- **Techniques for Knowledge Capture**:
  1. **Interviews**:
     - Conduct structured or semi-structured interviews with experts to document their insights and experiences.
     - Example: A senior engineer shares best practices for troubleshooting complex technical issues during an interview.

2. **Documentation**:
   - Use templates and structured forms to record processes, meeting notes, and project outcomes.
   - Example: A project manager documents lessons learned after the completion of a project.
3. **Observation and Shadowing**:
   - Observe experts in action to understand tacit knowledge embedded in their routines and decision-making processes.
   - Example: Junior employees shadow experienced sales professionals to learn negotiation techniques.

- **Tools for Knowledge Capture**:
  - **Knowledge Graphs**: Visual representations of relationships between concepts, data, and people.
  - **Video Recordings**: Capture demonstrations, lectures, or brainstorming sessions for future reference.
  - **Collaborative Platforms**: Tools like Notion and Evernote allow teams to document and organize knowledge collaboratively.
- **Challenges**:
  - Difficulty in articulating tacit knowledge.
  - Resistance from employees to share expertise due to time constraints or fear of losing their competitive edge.

2. Knowledge Organization

Definition: Knowledge organization focuses on structuring captured knowledge to make it easily accessible and retrievable. Effective organization ensures that employees can quickly find the information they need, reducing time spent searching for resources and improving productivity.

- **Techniques for Knowledge Organization**:
  1. **Taxonomies**:
     - Create hierarchical categorizations of knowledge based on topics, departments, or processes.
     - Example: A knowledge repository for a tech company might categorize content under headings like "Product Development," "Customer Support," and "Marketing."
  2. **Metadata**:
     - Add descriptive tags or labels to knowledge assets, enabling efficient search and retrieval.

- Example: Assigning keywords such as "sales strategy" or "market trends" to reports.

3. **Version Control**:
   - Implement systems to track changes in documents, ensuring that users access the latest version.
   - Example: A content management system (CMS) highlights updates to product manuals.

- **Tools for Knowledge Organization**:
  - **Content Management Systems (CMS)**:
    - Platforms like SharePoint and Drupal store, organize, and manage digital content.
  - **Enterprise Resource Planning (ERP) Software**:
    - Integrates knowledge organization into broader business processes.
    - Example: SAP ERP links operational data with knowledge assets.
  - **Document Management Systems**:
    - Systems like DocuWare and M-Files help classify and archive knowledge assets.
- **Challenges**:
  - Ensuring consistent categorization across large, dynamic datasets.
  - Managing the complexity of organizing unstructured data like images, videos, and informal notes.

3. Knowledge Sharing

Definition: Knowledge sharing involves disseminating organized knowledge across teams, departments, and individuals within an organization. This process fosters collaboration, innovation, and continuous learning by enabling employees to access and contribute to the organization's collective intelligence.

- **Techniques for Knowledge Sharing**:
  1. **Collaboration Tools**:
     - Platforms like Slack, Microsoft Teams, and Google Workspace facilitate communication and file sharing.
     - Example: Teams use shared Google Drive folders to collaborate on research and development projects.
  2. **Social Intranets**:
     - Internal platforms act as hubs for peer-to-peer interaction and resource sharing.

        - Example: Employees post questions, share updates, and access training materials on the company intranet.

    3. **Communities of Practice**:
        - Groups of employees with shared interests or expertise meet to exchange insights and solve problems collaboratively.
        - Example: A "Data Science Community" discusses machine learning techniques and shares code snippets.

- **Tools for Knowledge Sharing**:
    - **Wikis**:
        - Platforms like Confluence and MediaWiki enable employees to collaboratively create and edit knowledge repositories.
    - **Video Conferencing Software**:
        - Tools like Zoom and Webex support virtual knowledge-sharing sessions, including training and brainstorming.
    - **Learning Management Systems (LMS)**:
        - Systems like Moodle and Canvas deliver training modules and track employee learning progress.
- **Challenges**:
    - Encouraging employees to actively participate in knowledge-sharing initiatives.
    - Overcoming siloed communication within large organizations.

4. Knowledge Application

Definition: Knowledge application refers to the practical use of shared and organized knowledge in workflows, decision-making, and innovation. This process bridges the gap between theoretical knowledge and actionable insights, ensuring that the organization's intellectual resources are effectively utilized.

- **Techniques for Knowledge Application**:
    1. **AI-Driven Recommendations**:
        - Use artificial intelligence to provide contextual suggestions based on past projects, customer feedback, or market trends.
        - Example: An AI system recommends optimized inventory levels based on historical sales data.
    2. **Process Integration**:
        - Embed knowledge assets directly into workflows through automated tools and templates.
        - Example: A customer service representative accesses a troubleshooting guide during a live support call.

3. **Scenario Analysis:**
    - Apply historical knowledge to simulate potential outcomes and make informed decisions.
    - Example: A financial analyst uses past market data to predict stock performance.

- **Examples of Knowledge Application:**
    - A marketing team uses insights from a KM system to design a campaign tailored to customer preferences.
    - An R&D department builds on archived research to develop a new product line.
- **Tools for Knowledge Application:**
    - **Business Intelligence (BI) Tools:**
        - Platforms like Tableau and Power BI provide visual analytics and actionable insights.
    - **Project Management Software:**
        - Tools like Asana and Trello integrate knowledge into task management.
- **Challenges:**
    - Ensuring that employees apply knowledge consistently across projects.
    - Balancing the use of existing knowledge with the flexibility to innovate.

The Interconnected Nature of KM Processes

The core processes of knowledge management are interconnected, forming a continuous cycle:

1. **Knowledge Capture** initiates the process by collecting valuable insights and information.
2. **Knowledge Organization** ensures that captured knowledge is systematically structured for easy access.
3. **Knowledge Sharing** disseminates the organized knowledge across the organization.
4. **Knowledge Application** transforms shared knowledge into actionable insights, creating value for the organization.

Importance of Archiving

Archiving is a critical component of knowledge management and information preservation, ensuring that valuable records are maintained for future use. In a rapidly evolving digital world, where information is constantly generated and updated, the importance of archiving cannot be overstated. Archives serve as repositories of cultural, institutional, and historical memory, support compliance with regulatory frameworks, facilitate research and innovation, and provide safeguards against data loss.

1. Preserving History

One of the fundamental purposes of archiving is the preservation of history. Archives play a crucial role in safeguarding records of cultural, institutional, and societal importance, ensuring that future generations have access to the knowledge, traditions, and achievements of the past.

- **Cultural and Societal Significance**:
  - Archives maintain artifacts, manuscripts, photographs, and audiovisual materials that document cultural heritage and societal developments.
  - Example: National archives preserve government records, historical treaties, and cultural artifacts that reflect a nation's identity and history.
- **Institutional Memory**:
  - Organizations use archives to retain records of key decisions, policies, and milestones, ensuring continuity and institutional knowledge.
  - Example: Corporate archives store board meeting minutes, annual reports, and strategic documents to inform future leadership.
- **Global Importance**:
  - Archives contribute to preserving world history, enabling cross-cultural understanding and collaboration.
  - Example: The United Nations Archives and Records Management Section (ARMS) safeguards materials that document global efforts for peace, development, and human rights.
- **Challenges in Digital Preservation**:
  - The shift to digital formats poses challenges for preserving digital history, such as file format obsolescence and data degradation.
  - Solutions: Use of archival standards like PDF/A and regular migration of digital files to updated formats.

2. Ensuring Compliance

Archiving is essential for meeting legal, regulatory, and industry-specific requirements. Proper archiving practices help organizations avoid legal liabilities and maintain operational transparency.

- **Regulatory Requirements**:
  - Many industries, such as finance, healthcare, and education, are subject to strict regulations regarding record retention.
  - Example: Financial institutions are required to retain transaction data for several years to comply with auditing and anti-money laundering laws.
- **Transparency and Accountability**:
  - Archives ensure that records of decisions, contracts, and communications are available for auditing and accountability purposes.

    - Example: Government agencies archive public records, such as budgets and policies, to maintain transparency.
- **Legal Safeguards**:
    - In case of legal disputes, archived records serve as evidence to protect organizations and individuals.
    - Example: Employment records retained by HR departments provide documentation for resolving workplace disputes or verifying compliance with labor laws.
- **Emerging Regulations**:
    - Data protection laws, such as the General Data Protection Regulation (GDPR) and the California Consumer Privacy Act (CCPA), emphasize the importance of archiving practices that balance accessibility with privacy protections.
    - Organizations must implement secure archiving solutions to meet these requirements.

3. Supporting Research

Archives are indispensable resources for academic, scientific, and market research. They provide access to historical data, documented experiences, and curated collections that inform studies and foster innovation.

- **Academic Research**:
    - Universities and research institutions maintain digital repositories of theses, dissertations, and published papers to support scholarly activities.
    - Example: The Digital Public Library of America (DPLA) aggregates millions of digitized items from libraries, museums, and archives across the United States, enabling researchers to access a wealth of resources.
- **Scientific Discovery**:
    - Archived datasets, lab notes, and experimental results provide foundational materials for replicating studies and validating findings.
    - Example: Medical research institutions archive clinical trial data to support ongoing studies and regulatory reviews.
- **Market Insights**:
    - Companies use archived sales data, customer feedback, and industry reports to analyze trends and forecast future performance.
    - Example: Retailers analyze archived transaction records to understand seasonal shopping behaviors and optimize inventory.
- **Digital Accessibility**:
    - The digitization of archives has significantly expanded access to research materials, breaking down barriers of geography and physical availability.

- Example: Online platforms like JSTOR and ProQuest provide access to archived academic journals and historical documents.

4. Disaster Recovery

Archives play a vital role in disaster recovery, acting as backups to restore data and systems in the event of technical failures, cyberattacks, or natural disasters.

- **Data Resilience**:
  - Archived records ensure the availability of critical information even when primary systems are compromised.
  - Example: Cloud-based archives provide redundant storage solutions that protect against physical damage to servers.
- **Business Continuity**:
  - Archiving supports continuity planning by enabling organizations to quickly resume operations after disruptions.
  - Example: A financial firm relies on archived client data to restore account records following a ransomware attack.
- **Government Preparedness**:
  - Government archives ensure that essential records, such as birth certificates, land deeds, and legal documents, are preserved and accessible during emergencies.
  - Example: Emergency management agencies archive disaster response plans to streamline future recovery efforts.
- **Technological Safeguards**:
  - Advances in archival technology, such as distributed storage systems and blockchain, enhance the security and reliability of digital archives.
  - Example: Distributed ledger systems ensure that archived records remain tamper-proof and verifiable.

Broader Implications of Archiving

The importance of archiving extends beyond the four core areas described above, touching on ethical, economic, and environmental considerations.

- **Ethical Responsibility**:
  - Archives uphold the principles of transparency, accountability, and truth, preserving records that reflect the complexities of history and society.
  - Example: Archives documenting human rights abuses play a critical role in justice and reconciliation efforts.

- **Economic Value**:
    - Archives provide organizations with insights that drive innovation, enhance competitiveness, and support strategic planning.
    - Example: Historical marketing data helps businesses refine branding strategies and anticipate consumer trends.
- **Environmental Sustainability**:
    - Digitization reduces the need for physical storage, minimizing environmental impacts such as paper waste and energy consumption.
    - Example: Companies transitioning to paperless offices use digital archiving solutions to reduce their carbon footprint.

Types of Archiving

Archiving serves as a critical process for preserving knowledge and records across various domains, ranging from cultural heritage to business operations. The methods and systems used for archiving have evolved significantly over time, transitioning from traditional physical formats to advanced digital systems, with hybrid solutions often bridging the gap. Each type of archiving—**physical archives**, **digital archives**, and **hybrid archives**—offers unique advantages and limitations, tailored to specific needs and contexts.

1. Physical Archives

Definition: Physical archives refer to the traditional methods of storing tangible records such as paper documents, photographs, film reels, artifacts, and other physical materials. These archives are maintained in controlled environments to ensure the preservation of their content over time.

Characteristics of Physical Archives:

- **Storage Techniques**:
    - Documents and records are organized in filing cabinets, boxes, or shelves and often cataloged manually.
    - Photographs, artworks, and artifacts are stored in acid-free containers, with climate control systems maintaining optimal temperature and humidity levels.
- **Controlled Environments**:
    - Specialized facilities such as libraries, museums, and archival repositories are used to store materials safely.
    - Environmental controls protect materials from deterioration caused by factors like moisture, heat, and light exposure.
- **Applications**:
    - Historical preservation: National archives and museums house records of cultural and historical significance, such as treaties, manuscripts, and artifacts.
    - Organizational records: Companies store contracts, legal documents, and other vital records in secure, physical locations for compliance and reference.

Advantages:

1. **Tangible Records**:
   - Physical archives provide a concrete, visual, and tactile connection to historical artifacts and original documents.
   - Example: Historical manuscripts preserved in museums offer authentic, direct engagement with the past.
2. **Resilience to Cyber Threats**:
   - Physical records are immune to digital threats like hacking, ransomware, or data corruption.
3. **Cultural Significance**:
   - Original documents, photographs, and artifacts carry inherent historical and cultural value, making their physical preservation essential.

Limitations:

1. **Vulnerability to Physical Damage**:
   - Susceptible to disasters such as floods, fires, and pests.
   - Example: Archives destroyed during natural calamities like hurricanes or earthquakes.
2. **Limited Accessibility**:
   - Physical archives require individuals to visit specific locations to access materials, limiting their reach and usability.
3. **Space Requirements**:
   - Physical storage consumes significant space, and expanding facilities can be costly and impractical.
4. **Deterioration Over Time**:
   - Even with proper care, materials like paper and film degrade over decades.

2. Digital Archives

Definition: Digital archives store data in electronic formats, enabling efficient retrieval, duplication, and sharing. They leverage modern computing technologies to preserve and manage information with increased accessibility and scalability.

Characteristics of Digital Archives:

- **Storage Techniques**:
  - Data is stored in digital formats such as PDFs, JPEGs, or video files, often organized within databases or content management systems (CMS).
  - Cloud-based solutions, such as Amazon S3 and Google Cloud Storage, provide secure, scalable storage.

- **Accessibility**:
  - Users can access digital archives remotely via the internet, reducing geographic and logistical barriers.
  - Example: Academic institutions host open-access repositories, such as arXiv for scientific papers.
- **Automation**:
  - Metadata tagging and search algorithms enable automated indexing and retrieval of records.

Advantages:

1. **Ease of Access**:
   - Digital archives allow users to retrieve information instantly, from anywhere with an internet connection.
   - Example: Researchers can access digital collections like JSTOR and Project MUSE without visiting physical libraries.
2. **Scalability**:
   - Digital storage can grow dynamically to accommodate increasing volumes of data without the need for physical expansion.
3. **Cost Efficiency**:
   - Reduces expenses related to physical storage, such as real estate, maintenance, and manual labor.
4. **Enhanced Preservation**:
   - Digital formats eliminate the risks of physical wear and tear, offering long-term reliability.
5. **Searchability**:
   - Advanced search tools enable users to locate specific files or information quickly, improving productivity.

Challenges:

1. **Cybersecurity Risks**:
   - Digital archives are vulnerable to hacking, data breaches, and malware attacks.
   - Example: Cyberattacks on government servers threaten the integrity of sensitive records.
2. **Data Obsolescence**:
   - File formats and storage media can become outdated, necessitating periodic migration to newer technologies.

3. **Initial Costs**:
    - While long-term savings are significant, establishing a secure and efficient digital archive requires substantial initial investment in technology and infrastructure.
4. **Digital Divide**:
    - Limited internet access or technological expertise can restrict the usability of digital archives for certain populations or regions.

3. Hybrid Archives

Definition: Hybrid archives combine physical and digital methods to create a comprehensive system of preservation that leverages the strengths of both approaches. This dual strategy ensures redundancy, accessibility, and flexibility.

Characteristics of Hybrid Archives:

- **Integration**:
    - Hybrid archives maintain physical records while digitizing them for enhanced accessibility and backup.
    - Example: Libraries that store rare manuscripts physically but provide digital copies for public access.
- **Redundancy**:
    - Critical records are preserved in multiple formats to minimize risks of loss or damage.
    - Example: Governments archive vital records in both physical vaults and cloud-based systems.
- **Flexibility**:
    - Users can choose the format that best suits their needs, whether physical authenticity or digital convenience.

Advantages:

1. **Comprehensive Preservation**:
    - Physical and digital formats complement each other, ensuring that records are preserved under various contingencies.
2. **Enhanced Accessibility**:
    - Digitized records expand the reach of physical archives, making them available to global audiences.
3. **Risk Mitigation**:
    - Dual formats protect against risks unique to each medium, such as natural disasters for physical archives and cyberattacks for digital archives.

4. **Cost Balance**:
    - o Organizations can prioritize digitization for high-demand records while maintaining physical copies for archival purposes.

Challenges:

1. **Complexity**:
    - o Managing both physical and digital systems requires significant coordination and resource allocation.
2. **Cost Implications**:
    - o Hybrid solutions can be more expensive to implement and maintain than single-format archives.
3. **Consistency Issues**:
    - o Ensuring that physical and digital records remain synchronized can be challenging.

Technologies in Knowledge Management and Archiving

Digital transformation has fundamentally reshaped the landscape of knowledge management (KM) and archiving. Advanced technologies such as Artificial Intelligence (AI), cloud computing, blockchain, Optical Character Recognition (OCR), and big data analytics have revolutionized the way knowledge is captured, organized, shared, and preserved. These technologies enhance the scalability, efficiency, and security of KM and archiving systems, enabling organizations to harness their intellectual and informational assets more effectively.

1. Artificial Intelligence (AI) and Machine Learning (ML)

AI and ML are at the forefront of the digital revolution in KM and archiving, offering automation and intelligence to handle complex processes such as data categorization, content recommendations, and information retrieval.

Applications:

- **Knowledge Categorization**:
    - o AI systems classify documents, emails, and other forms of data based on predefined taxonomies, improving organization and searchability.
    - o Example: An AI algorithm tags legal contracts by type, jurisdiction, and expiration date, streamlining document retrieval.
- **Content Recommendations**:
    - o ML algorithms analyze user behavior and historical data to provide personalized suggestions for relevant documents or resources.
    - o Example: A KM platform suggests project templates or case studies based on a user's past queries.

- **Chatbots and Virtual Assistants**:
  - AI-powered chatbots enable users to interact with KM systems conversationally, accessing relevant information quickly.
  - Example: A customer service chatbot retrieves troubleshooting guides from an archived knowledge base to assist users.

Benefits:

- **Efficiency**:
  - Automates repetitive tasks, reducing manual effort in organizing and retrieving data.
- **Scalability**:
  - Handles large volumes of data, adapting to the growing information needs of organizations.
- **Real-Time Insights**:
  - Enables dynamic analysis and decision-making through predictive analytics.

Challenges:

- **Data Bias**:
  - ML models can inherit biases from training data, affecting accuracy and fairness.
- **Integration Complexity**:
  - Implementing AI systems requires robust infrastructure and expertise.

2. Cloud Computing

Cloud computing has emerged as a cornerstone of modern KM and archiving, providing flexible, scalable, and cost-effective solutions for data storage and management.

Applications:

- **Scalable Storage**:
  - Cloud platforms offer on-demand storage that scales with organizational needs, eliminating the constraints of physical infrastructure.
  - Example: Google Drive and Microsoft OneDrive allow users to store, share, and access files collaboratively.
- **Disaster Recovery**:
  - Cloud services provide backup and redundancy, ensuring data resilience in case of technical failures or disasters.
  - Example: Organizations use Amazon Web Services (AWS) to replicate critical archives across multiple geographic locations.

- **Global Accessibility**:
  - Cloud-based KM systems enable remote teams to access knowledge repositories from anywhere with an internet connection.

Benefits:

- **Cost-Effectiveness**:
  - Reduces capital expenditure on physical storage infrastructure.
- **Flexibility**:
  - Supports remote work and real-time collaboration.
- **Security**:
  - Offers advanced encryption and access control features to protect sensitive data.

Challenges:

- **Data Privacy**:
  - Storing sensitive information on third-party platforms raises concerns about compliance with regulations like GDPR.
- **Dependency on Connectivity**:
  - Cloud services require stable internet access, which can be a limitation in remote areas.

3. Blockchain Technology

Blockchain offers a decentralized, tamper-proof framework for secure data management and archiving, ensuring the integrity and authenticity of records.

Applications:

- **Data Integrity**:
  - Blockchain creates immutable records, making it ideal for archiving critical documents such as legal contracts, financial records, and intellectual property.
  - Example: A law firm uses blockchain to archive digital contracts, ensuring they remain unaltered over time.
- **Provenance Tracking**:
  - Ensures transparency by tracking the history and ownership of archived data.
  - Example: Museums use blockchain to document the provenance of art pieces and historical artifacts.
- **Decentralized Storage**:
  - Blockchain-based storage solutions distribute data across nodes, enhancing redundancy and security.

Benefits:

- **Security**:
  - Resistant to tampering and unauthorized modifications.
- **Transparency**:
  - Maintains a verifiable audit trail for archived records.
- **Decentralization**:
  - Reduces reliance on centralized storage systems, mitigating risks of single points of failure.

Challenges:

- **Energy Consumption**:
  - Blockchain networks, especially those using proof-of-work mechanisms, consume significant energy.
- **Scalability**:
  - Limited transaction processing speed can hinder the adoption of blockchain for large-scale archiving.

4. Optical Character Recognition (OCR)

OCR technology bridges the gap between physical and digital archives by converting scanned documents into machine-readable formats, enabling easier storage, search, and retrieval.

Applications:

- **Document Digitization**:
  - OCR transforms printed and handwritten text into digital text, creating searchable digital copies of physical records.
  - Example: Libraries digitize historical manuscripts using OCR to make them accessible online.
- **Data Extraction**:
  - Extracts specific information, such as names or dates, from scanned documents for indexing and analysis.
  - Example: Financial institutions use OCR to process invoices and receipts.
- **Accessibility Enhancement**:
  - Converts printed materials into formats compatible with screen readers, improving accessibility for visually impaired users.

Benefits:

- **Efficiency**:
    - Automates the labor-intensive process of manual data entry.
- **Preservation**:
    - Protects fragile physical documents by reducing handling during access.
- **Searchability**:
    - Enables keyword searches within digitized documents.

Challenges:

- **Accuracy**:
    - OCR struggles with poor-quality scans or complex layouts, requiring manual validation.
- **Language Support**:
    - Limited capability to process texts in less common languages or scripts.

5. Big Data Analytics

Big data analytics leverages advanced computational techniques to analyze large datasets, uncovering patterns, trends, and actionable insights that enhance KM and archiving systems.

Applications:

- **Trend Analysis**:
    - Identifies patterns in archived data to forecast future developments.
    - Example: Retail companies analyze archived sales data to predict seasonal demand.
- **Customer Insights**:
    - Analyzes feedback and interaction logs to improve customer experiences.
    - Example: A telecommunications provider reviews archived customer complaints to optimize service delivery.
- **Operational Optimization**:
    - Analyzes historical project data to improve workflows and resource allocation.
    - Example: Construction firms use big data to predict project timelines based on archived performance metrics.

Benefits:

- **Decision-Making**:
    - Provides data-driven insights to inform strategic choices.

- **Scalability**:
    - Handles massive datasets efficiently, supporting organizations of all sizes.
- **Competitive Advantage**:
    - Enables proactive responses to market trends and emerging opportunities.

Challenges:

- **Complexity**:
    - Requires expertise in data science and analytics to extract meaningful insights.
- **Data Privacy**:
    - Analyzing sensitive data necessitates compliance with privacy regulations.

Challenges in Knowledge Management and Archiving

While Knowledge Management (KM) and archiving offer transformative benefits in preserving and utilizing information, they also face significant challenges. As the volume, complexity, and sensitivity of information continue to grow, organizations must navigate obstacles related to data management, security, interoperability, cost, and obsolescence. Addressing these challenges requires innovative approaches and the integration of advanced technologies to ensure the sustainability and effectiveness of KM and archiving systems.

1. Data Overload

Challenge: The exponential growth of digital information has created an environment of data overload, where managing, organizing, and archiving massive volumes of information becomes increasingly difficult. With data generated from multiple sources—such as emails, social media, IoT devices, and operational records—organizations often struggle to differentiate valuable knowledge from redundant or irrelevant information.

Implications:

- Difficulty in identifying and prioritizing critical data for archiving.
- Increased storage requirements leading to higher costs.
- Overwhelmed KM systems result in inefficiencies and reduced productivity.

Solution:

AI-Driven Tools:

- **Automated Filtering**: Artificial Intelligence (AI) tools can analyze large datasets to identify duplicates, categorize content, and prioritize critical information for archiving.
    - Example: An AI-powered KM system automatically identifies high-value documents, such as contracts or research findings, for long-term preservation.
- **Data Deduplication**: Reduces storage needs by identifying and removing redundant files.

- **Predictive Analysis**: Machine Learning (ML) algorithms predict which information is most likely to be used in the future, enabling better archiving decisions.

2. Privacy and Security

Challenge: Sensitive information stored in KM systems and archives is vulnerable to unauthorized access, data breaches, and cyberattacks. As organizations increasingly rely on digital systems, ensuring data privacy and security becomes a critical concern, especially with the rise of stringent data protection regulations like GDPR and HIPAA.

Implications:

- Risk of financial losses and reputational damage due to data breaches.
- Legal liabilities arising from non-compliance with privacy regulations.
- Erosion of trust among stakeholders, including customers and employees.

Solution:

Robust Security Measures:

- **Encryption**: Encrypting archived data ensures that even if it is accessed without authorization, it remains unreadable.
- **Access Controls**: Role-based access ensures that only authorized personnel can access specific data.
    - Example: A financial institution implements multi-factor authentication for accessing sensitive archives.
- **Data Masking**: Protects sensitive data by obfuscating identifiers in archived datasets while maintaining usability.
- **Compliance Tools**: Use of tools to automate compliance checks and ensure adherence to regulations like GDPR.

3. Interoperability Issues

Challenge: The use of multiple systems and formats across departments or organizations often leads to interoperability issues, hindering seamless knowledge sharing and archiving. Proprietary formats, incompatible software, and siloed data repositories exacerbate the problem.

Implications:

- Reduced efficiency due to time lost in converting or transferring data.
- Increased costs associated with integrating incompatible systems.
- Fragmentation of knowledge, limiting its accessibility and usability.

Solution:

Adoption of Open Standards and APIs:

- **Open Standards**: Utilize widely accepted formats and protocols to ensure compatibility across systems.
  - Example: Archiving data in non-proprietary formats like PDF/A, XML, or CSV ensures long-term accessibility.
- **APIs (Application Programming Interfaces)**: APIs facilitate seamless integration between different systems, enabling data exchange and collaboration.
  - Example: An API connects a customer relationship management (CRM) system with a KM platform, ensuring that customer insights are archived automatically.

4. Cost

Challenge: Implementing and maintaining KM and archiving systems can be expensive, particularly for small and medium-sized organizations. Costs include hardware, software, storage, maintenance, and employee training.

Implications:

- Financial strain on organizations with limited budgets.
- Underinvestment in KM and archiving systems leads to inefficiencies and data loss.
- Difficulty in scaling systems to meet growing data demands.

Solution:

Cloud-Based Services:

- **Scalability**: Cloud storage solutions allow organizations to scale storage needs dynamically, paying only for what they use.
- **Cost Efficiency**: Eliminates the need for significant upfront investment in hardware and reduces ongoing maintenance costs.
  - Example: A startup uses Google Workspace to archive emails, documents, and collaboration records, avoiding the need for in-house servers.
- **Subscription Models**: Flexible pricing models make advanced archiving tools accessible to smaller organizations.

5. Obsolescence

Challenge: Digital archives risk becoming inaccessible due to outdated software, hardware, or file formats. As technology evolves, older systems may no longer be supported, making it challenging to retrieve archived data.

Implications:

- Loss of critical information stored in obsolete formats.
- High costs associated with migrating data to modern platforms.
- Disruption of workflows due to compatibility issues.

Solution:

Regular Updates and Migration:

- **Format Migration:** Periodically convert archived data to current, widely supported formats to ensure accessibility.
    - Example: A library migrates digitized manuscripts from outdated TIFF formats to more versatile PDFs.
- **Hardware Upgrades:** Replace aging storage devices with modern alternatives to avoid data loss due to hardware failure.
- **Backward Compatibility:** Use systems that support legacy formats while offering modern features.
- **Emulation:** Deploy software that emulates older environments to access data stored in outdated formats.

The Role of New Media in Knowledge Management and Archiving

New media has revolutionized the field of knowledge management (KM) and archiving by introducing interactivity, user-generated content, and global connectivity. Unlike traditional methods, which often rely on static, centralized processes, new media provides dynamic and participatory approaches to creating, managing, and preserving knowledge. Platforms like social media, collaborative tools, and digital repositories have transformed how organizations and societies collect, share, and retain information.

1. Crowdsourced Knowledge

Definition: Crowdsourcing in KM leverages collective contributions from individuals or communities to create, organize, and manage knowledge. This approach decentralizes the knowledge creation process, empowering a diverse range of participants to contribute.

Applications:

- **Collaborative Platforms:**
    - Platforms like Wikipedia, GitHub, and Stack Overflow rely on user contributions to build vast knowledge repositories.
    - Example: Wikipedia is a globally recognized crowdsourced encyclopedia, with millions of articles created and edited by volunteers worldwide.

- **Open Data Initiatives**:
    - Crowdsourcing is used in scientific and governmental projects to gather data or solve problems.
    - Example: Citizen science projects, such as Galaxy Zoo, allow the public to assist in classifying astronomical data.

Advantages:

- **Diversity of Perspectives**:
    - Crowdsourcing includes contributions from people of varied backgrounds, ensuring a broader and more inclusive knowledge base.
    - Example: Wikipedia articles often present information from multiple cultural and historical viewpoints.
- **Real-Time Updates**:
    - User-generated content can be updated instantly, ensuring that information remains current.
    - Example: Breaking news on social media platforms like Twitter is often faster than traditional media channels.

Challenges:

- **Quality Control**:
    - The open nature of crowdsourced platforms can lead to inaccuracies or biased content.
    - Solution: Employing moderation, peer review, and verification processes to ensure reliability.
- **Ownership and Attribution**:
    - Defining intellectual property rights for crowdsourced content can be complex.

Impact: Crowdsourced knowledge democratizes information creation, making it more accessible and representative while fostering collaboration and innovation.

2. Social Media as Archives

Definition: Social media platforms serve as informal archives that document cultural, social, and political trends in real time. They capture vast amounts of data, including text, images, videos, and user interactions, making them valuable resources for studying societal dynamics.

Applications:

- **Cultural and Social Trends**:
    - Social media preserves snapshots of public sentiment, cultural expressions, and major events.

  - Example: The #MeToo and #BlackLivesMatter movements are well-documented through social media platforms, serving as critical archives of activism.
- **Historical Events**:
  - Platforms like Twitter and Instagram have become primary sources for documenting real-time events, from natural disasters to political revolutions.
  - Example: The Arab Spring was extensively chronicled through tweets, posts, and shared videos.

Advantages:

- **Global Reach**:
  - Social media transcends geographic boundaries, capturing diverse voices and perspectives.
- **Accessibility**:
  - Archived social media content is easily searchable and can be analyzed to understand societal behaviors and trends.

Challenges:

- **Misinformation**:
  - The prevalence of fake news and unverified content on social media poses a challenge to its reliability as an archival source.
  - Solution: Implementing fact-checking mechanisms and labeling unverified content.
- **Data Permanence**:
  - Posts, tweets, and videos can be deleted or lost due to account closures, policy changes, or platform shutdowns.
  - Solution: Initiatives like the Internet Archive's Wayback Machine aim to preserve web-based content, including social media data.

Impact: Social media archives offer a unique lens to study contemporary culture and history, but their ephemeral nature underscores the need for formal preservation strategies.

3. Multimedia Integration

Definition: New media enables the archiving of diverse content types, including text, images, audio, and video. This multimedia approach enriches KM systems by preserving knowledge in various formats, catering to different user needs and preferences.

Applications:

- **Digital Storytelling**:
  - Platforms like StoryCorps and YouTube preserve oral histories, cultural traditions, and personal narratives in audio and video formats.

    - Example: StoryCorps archives personal stories to document the diverse experiences of individuals across the United States.
- **Visual Archives**:
    - Image and video-sharing platforms like Flickr and Vimeo serve as repositories for visual content, enabling communities to share and preserve their work.
    - Example: Museums digitize artwork and create online galleries for public access.
- **Interactive Archives**:
    - Virtual and augmented reality (VR/AR) technologies create immersive archives, such as virtual museums or 3D reconstructions of historical sites.
    - Example: Google Arts & Culture allows users to virtually explore museums and artifacts from around the world.

Advantages:

- **Enhanced Accessibility**:
    - Multimedia content caters to diverse audiences, including those with varying literacy levels or learning preferences.
    - Example: Educational videos make complex concepts accessible to a broader audience.
- **Engagement**:
    - Interactive and multimedia archives engage users more effectively than static records.
- **Preservation of Diverse Content**:
    - Multimedia integration ensures that non-textual knowledge, such as oral traditions or visual art, is archived for future generations.

Challenges:

- **Storage and Bandwidth**:
    - Multimedia content requires significant storage space and bandwidth, which can be costly for large-scale archives.
    - Solution: Cloud-based storage solutions like AWS and Google Cloud offer scalable options.
- **Format Obsolescence**:
    - Digital formats for multimedia content can become outdated, requiring periodic migration to ensure accessibility.
    - Solution: Use open and widely supported formats like MP4 for video and JPEG for images.

Impact: Multimedia integration broadens the scope of KM and archiving, ensuring that diverse forms of knowledge are preserved and made accessible to global audiences.

Networks

Networks are the backbone of modern communication and information systems, enabling devices, systems, and people to connect, share, and interact seamlessly. In the context of Knowledge Management (KM) and archiving, networks play a critical role in facilitating the efficient transfer, storage, and retrieval of information. From local area networks (LANs) in offices to global internet-based infrastructures, networks provide the necessary framework for collaboration, accessibility, and data integration across platforms.

Definition of Networks

A network is a system of interconnected devices or nodes that exchange data and resources using communication protocols. Networks enable the transfer of information, allowing individuals and organizations to collaborate, access shared knowledge, and store data efficiently.

Types of Networks

Networks can be categorized based on their scale, architecture, and functionality. The following are the primary types of networks that support KM and archiving:

1. **Local Area Networks (LANs)**:
    - **Definition**: LANs connect devices within a limited geographic area, such as an office, school, or home.
    - **Application**:
        - Organizations use LANs to enable employees to share files, access common resources, and connect to printers or storage devices.
        - Example: A corporate KM system hosted on an internal LAN provides employees access to shared knowledge repositories.
    - **Advantages**:
        - High-speed data transfer.
        - Enhanced security due to restricted access.
    - **Limitations**:
        - Limited to a specific physical location.
2. **Wide Area Networks (WANs)**:
    - **Definition**: WANs connect multiple LANs over large geographic areas, often spanning cities, countries, or continents.
    - **Application**:
        - Global organizations use WANs to connect branch offices, enabling centralized KM systems and archiving solutions.
        - Example: An international company uses a WAN to provide employees in different locations access to a centralized document management system.

- **Advantages**:
  - Facilitates global connectivity and data sharing.
- **Limitations**:
  - Higher latency and cost compared to LANs.

3. **Wireless Networks**:
   - **Definition**: Wireless networks use radio signals, infrared, or satellite communication to connect devices without physical cables.
   - **Application**:
     - Mobile access to KM platforms and archives.
     - Example: Employees use Wi-Fi to access a cloud-based KM platform from their smartphones during a meeting.
   - **Advantages**:
     - Provides mobility and flexibility.
   - **Limitations**:
     - Vulnerable to security threats like hacking and unauthorized access.
4. **The Internet**:
   - **Definition**: The global network of interconnected devices that facilitates the sharing of information and resources.
   - **Application**:
     - The internet is essential for accessing cloud-based KM systems, digital archives, and collaborative tools.
     - Example: A university accesses digital repositories and research databases hosted online.
   - **Advantages**:
     - Universal access and scalability.
   - **Limitations**:
     - Dependence on reliable connectivity and cybersecurity risks.
5. **Cloud Networks**:
   - **Definition**: Networks that leverage cloud infrastructure to store, manage, and access data over the internet.
   - **Application**:
     - Cloud networks host KM systems and archives, offering remote accessibility and scalability.

- Example: An organization uses Microsoft Azure to store and manage its knowledge repositories.

- **Advantages**:
  - On-demand scalability and cost efficiency.
- **Limitations**:
  - Potential concerns regarding data privacy and vendor dependency.

Role of Networks in KM and Archiving

Networks are integral to the success of KM and archiving systems, facilitating seamless communication, collaboration, and data management. Their role includes:

1. **Data Sharing**:
   - Networks enable real-time sharing of documents, records, and knowledge across devices and locations.
   - Example: A collaborative team accesses shared knowledge repositories via a VPN (Virtual Private Network).
2. **Centralized Storage and Retrieval**:
   - Networks allow centralized KM systems and archives to be accessed by authorized users from different locations.
   - Example: Employees in different time zones collaborate on a project using a cloud-based KM platform.
3. **Collaboration and Communication**:
   - Networks support tools like video conferencing, chat applications, and shared workspaces, fostering team collaboration.
   - Example: Teams use platforms like Slack or Microsoft Teams to share updates and knowledge resources.
4. **Security and Access Control**:
   - Advanced networking technologies enable secure access to KM and archiving systems through encryption, firewalls, and authentication protocols.
   - Example: An archive accessible via a secure VPN ensures data integrity and privacy.
5. **Scalability and Flexibility**:
   - Networks provide the infrastructure for scaling KM systems and archives as organizational needs grow.
   - Example: A startup expands its KM capabilities by integrating cloud-based tools as its workforce grows.

Emerging Networking Technologies in KM and Archiving

The rapid evolution of networking technologies has introduced new possibilities for KM and archiving systems:

1. **5G Networks**:
   - **Definition**: The fifth generation of mobile networks offering ultra-fast speeds and low latency.
   - **Impact**:
     - Supports real-time data sharing and access to cloud-based archives from mobile devices.
     - Enables immersive experiences such as augmented reality (AR) and virtual reality (VR) in KM applications.
2. **Edge Computing**:
   - **Definition**: Processes data closer to its source rather than relying on centralized servers.
   - **Impact**:
     - Reduces latency for KM systems, enabling faster access to archives and real-time analytics.
     - Example: IoT devices at remote locations use edge computing to sync data with centralized KM platforms.
3. **Software-Defined Networking (SDN)**:
   - **Definition**: Allows network administrators to manage and optimize network performance programmatically.
   - **Impact**:
     - Enhances the efficiency and adaptability of networks supporting KM and archiving systems.
     - Example: SDN dynamically allocates bandwidth for a video archive during peak usage hours.
4. **Blockchain Networks**:
   - **Definition**: A decentralized system for storing data with cryptographic security.
   - **Impact**:
     - Ensures the integrity and authenticity of archived records, preventing tampering.
     - Example: Legal firms use blockchain to archive contracts and ensure traceability.
5. **Artificial Intelligence in Networking**:
   - **Definition**: AI-driven solutions optimize network management and performance.

- **Impact**:
  - Automates monitoring and troubleshooting for KM systems, ensuring uninterrupted access.
  - Example: An AI-powered network detects and resolves connectivity issues affecting an organization's knowledge repository.

Challenges in Networking for KM and Archiving

Despite their benefits, networks face challenges that can impact their efficiency in supporting KM and archiving systems:

1. **Bandwidth Limitations**:
   - Large datasets, such as multimedia archives, require substantial bandwidth, which may strain networks.
2. **Cybersecurity Risks**:
   - Networks are vulnerable to attacks such as phishing, ransomware, and data breaches, threatening KM and archive integrity.
3. **Latency and Downtime**:
   - Network delays or outages can disrupt access to critical KM resources.
4. **Integration Complexity**:
   - Connecting diverse systems and ensuring interoperability across networks can be challenging.

Social Media – Advantages and Limitations

Social media has emerged as one of the most transformative aspects of the digital age, redefining how people connect, communicate, and access information. From individuals seeking entertainment and networking opportunities to businesses leveraging it for marketing and brand-building, social media platforms have become integral to modern life. Platforms such as Facebook, Instagram, Twitter, LinkedIn, TikTok, and others have not only facilitated global interactions but also played a pivotal role in areas like knowledge dissemination, education, activism, and commerce. Despite its wide-ranging advantages, social media is not without its challenges. Issues such as misinformation, data privacy concerns, mental health impacts, and content moderation failures raise questions about its ethical implications and societal consequences.

Advantages of Social Media

Social media's rapid adoption and pervasive influence stem from its numerous benefits, which have reshaped communication, collaboration, and access to information.

1. Enhanced Connectivity

Social media platforms have broken down geographical and cultural barriers, enabling people to connect and interact on an unprecedented scale.

- **Global Reach**:
    - Platforms like Facebook, Instagram, and LinkedIn allow individuals, communities, and organizations to interact across continents, fostering personal relationships and professional collaborations.
    - Example: Families separated by distance use platforms like WhatsApp and Facebook Messenger to stay in touch through real-time video calls and messages.
- **Breaking Barriers**:
    - Social media has empowered individuals in remote and underprivileged regions to access global conversations and opportunities.
    - Example: Activists in developing countries use Twitter to raise awareness about social justice issues, gaining support from international communities.
- **Professional Networking**:
    - Platforms like LinkedIn connect professionals, enabling them to build networks, share knowledge, and explore career opportunities.

2. Democratization of Information

Social media has revolutionized how information is created, shared, and consumed, putting power into the hands of individuals rather than institutions.

- **User-Generated Content**:
    - Anyone with an internet connection can create and share content, bypassing traditional media gatekeepers.
    - Example: Bloggers and vloggers use platforms like YouTube and TikTok to share their expertise, ranging from cooking tutorials to financial advice.
- **Real-Time Updates**:
    - Social media enables instant dissemination of information, making it a critical tool for breaking news and live updates.
    - Example: During emergencies like natural disasters, social media platforms provide real-time updates about affected areas, relief measures, and emergency contacts.
- **Amplifying Marginalized Voices**:
    - Social media platforms amplify the voices of underrepresented groups, fostering inclusivity and awareness.
    - Example: Movements like #MeToo and #BlackLivesMatter gained global traction through the collective power of social media users.

3. Knowledge Sharing and Collaboration

Social media fosters collaboration and knowledge sharing across industries, disciplines, and communities.

- **Educational Content**:
  - Platforms like YouTube, LinkedIn Learning, and Instagram Reels provide valuable educational resources in accessible formats.
  - Example: Educational influencers create content about science, history, and personal development, reaching millions of viewers.
- **Crowdsourcing and Open Knowledge**:
  - Collaborative platforms like Reddit and Stack Overflow enable users to ask questions, share insights, and solve problems collectively.
  - Example: Programmers use GitHub to collaborate on coding projects and share knowledge.
- **Cross-Border Collaboration**:
  - Teams in different countries collaborate on projects using social media tools, enhancing global innovation.
  - Example: Scientists use Twitter to share research updates and engage in academic discussions.

4. Marketing and Business Growth

Social media has become an indispensable tool for businesses, enabling them to reach, engage, and convert customers.

- **Targeted Advertising**:
  - Platforms like Facebook and Instagram offer highly targeted advertising, allowing businesses to reach specific demographics based on interests, location, and behavior.
  - Example: An e-commerce company uses Instagram ads to target young adults interested in sustainable fashion.
- **Brand Visibility**:
  - Consistent social media activity builds brand awareness and loyalty, helping businesses establish a strong online presence.
  - Example: Small businesses use platforms like TikTok to create viral content that drives traffic to their websites.
- **Customer Engagement**:
  - Social media facilitates direct interaction with customers, enabling businesses to address concerns, gather feedback, and build relationships.
  - Example: Brands use Twitter to respond to customer complaints and queries in real time.

5. Community Building

Social media fosters the creation and growth of online communities centered on shared interests, goals, or identities.

- **Interest-Based Groups**:
    - Platforms like Facebook and Reddit host groups and subreddits dedicated to specific topics, from gardening to cryptocurrency.
    - Example: Hobbyists join photography groups on Facebook to share tips, critique each other's work, and find inspiration.
- **Support Networks**:
    - Social media provides a space for individuals to find emotional support and solidarity in difficult times.
    - Example: Mental health advocates use Instagram to share coping strategies and promote awareness about mental well-being.

6. Facilitating Activism and Social Change

Social media has emerged as a powerful tool for activism, enabling grassroots movements and global campaigns.

- **Awareness Campaigns**:
    - Activists use platforms to spread awareness about social, political, and environmental issues.
    - Example: Climate activists use hashtags like #FridaysForFuture to promote environmental sustainability.
- **Mobilizing Action**:
    - Social media facilitates the organization of protests, fundraisers, and petitions.
    - Example: Online petitions on platforms like Change.org are often promoted through social media to gather signatures and support.

Limitations of Social Media

Despite its transformative impact, social media poses several challenges that can undermine its benefits if not managed effectively.

1. Spread of Misinformation

Social media's speed and reach make it a breeding ground for the rapid dissemination of false information.

- **Echo Chambers**:
    - Algorithms prioritize content that aligns with users' beliefs, reinforcing biases and spreading misinformation.

  - Example: Political misinformation spreads rapidly within ideologically similar groups.
- **Fake News**:
  - Unverified and sensational content often garners more engagement than factual news.
  - Example: During the COVID-19 pandemic, false claims about vaccines proliferated on platforms like Facebook and WhatsApp.

2. Privacy and Security Concerns

Social media's reliance on user data raises ethical concerns about privacy and security.

- **Data Exploitation**:
  - Platforms collect and monetize personal data without transparent consent mechanisms.
  - Example: The Cambridge Analytica scandal revealed the misuse of Facebook user data for political manipulation.
- **Cybersecurity Risks**:
  - Accounts are vulnerable to hacking, identity theft, and phishing attacks.
  - Example: High-profile Twitter accounts were hacked in 2020 to promote cryptocurrency scams.

3. Mental Health Impacts

Excessive use of social media has been linked to negative mental health outcomes, particularly among younger users.

- **Comparison Culture**:
  - Curated content on platforms fosters unrealistic expectations and feelings of inadequacy.
  - Example: Users comparing their lives to influencers' highlight reels may experience anxiety or depression.
- **Addictive Design**:
  - Features like infinite scrolling and notifications are designed to maximize engagement, leading to addictive behaviors.
  - Example: Users spend hours on TikTok without realizing how much time has passed.

4. Cyberbullying and Harassment

The anonymity and reach of social media facilitate harmful behaviors such as cyberbullying and harassment.

- **Trolling and Abuse**:
  - Anonymous users target individuals with hateful or threatening messages.

- Example: Public figures and activists are often subjected to coordinated harassment campaigns.

- **Social Pressure**:
  - Teens and young adults face pressure to conform to trends, often leading to negative self-esteem.

5. Content Moderation Challenges

Social media platforms face difficulties in effectively moderating content, leading to the spread of harmful material.

- **Hate Speech**:
  - Algorithms struggle to identify and remove hate speech promptly.
  - Example: Delays in moderating harmful content can exacerbate social tensions.
- **Algorithm Bias**:
  - Automated systems may prioritize sensational content over meaningful discourse.

6. Economic Inequalities and Digital Divide

Not everyone has equal access to social media, perpetuating digital divides between different socioeconomic groups.

- **Cost of Access**:
  - Internet connectivity and devices remain out of reach for many in low-income regions.
  - Example: Limited internet penetration in rural areas restricts social media adoption.
- **Monetization Gaps**:
  - Wealthier creators and organizations often have more resources to succeed on platforms, leaving smaller players at a disadvantage.

Balancing the Pros and Cons

Social media's advantages in connectivity, information sharing, and community building are undeniable, but its limitations underscore the need for responsible use and regulation. Addressing these challenges requires collaborative efforts from platforms, policymakers, and users to create a safer and more equitable online environment.

Chapter 6

Digital Audio and Videography

Introduction

Multimedia, as a concept and practice, has profoundly transformed how individuals and societies communicate, create, and consume information. It represents the harmonious integration of diverse media types—text, images, audio, video, and interactivity—to craft immersive and impactful experiences. Over the decades, multimedia has transcended its role as a mere technological innovation, becoming a cultural force that bridges gaps between art, education, entertainment, and business. The evolution of multimedia is a testament to humanity's relentless pursuit of better tools for expression, storytelling, and interaction, driven by advances in technology and a deepening understanding of the human experience.

The journey of multimedia begins with its conceptual roots in traditional communication forms, where visual and auditory elements were combined to convey ideas and emotions. Long before the term "multimedia" was coined, cave paintings, theatrical performances, and oral storytelling embodied its essence by merging visual, verbal, and performative elements to engage audiences. These early forms of multimedia laid the groundwork for what would become a sophisticated interplay of various media formats, evolving alongside technological advancements. As societies advanced, so too did the tools and mediums used for communication. The invention of printing technology in the 15th century marked a significant leap in the dissemination of information, allowing for the integration of text and imagery on a scale previously unimaginable. Illustrated books, pamphlets, and posters became early examples of static multimedia, where visual and textual elements coexisted to enhance understanding and appeal. Similarly, the rise of theatrical arts introduced a dynamic interplay of spoken word, music, and stage visuals, reflecting multimedia's growing potential for immersion and narrative depth.

The 20th century heralded a new era of multimedia with the advent of analog technologies. Photography and cinema emerged as powerful tools for storytelling, offering a blend of static and moving visuals. Silent films, often accompanied by live musical performances, demonstrated the potential of combining audio and visual elements to evoke emotions and tell compelling stories. The subsequent introduction of synchronized sound revolutionized cinema, marking the beginning of modern audiovisual media. Simultaneously, the invention of radio and television brought audio and video into households, reshaping how people consumed entertainment and news. These innovations established the foundations of multimedia as a vehicle for mass communication, accessibility, and shared cultural experiences. The transition from analog to digital technologies in the late 20th century marked a turning point in the evolution of multimedia. The advent of personal computers and digital devices revolutionized how multimedia content was created, distributed, and consumed. Digital tools

DOI: 10.4324/9781003750444-6

enabled the seamless integration of text, images, audio, and video, breaking down the barriers between different media formats. Compact discs (CDs) and digital versatile discs (DVDs) offered new ways to store and share multimedia content, while software applications like Adobe Photoshop and Microsoft PowerPoint empowered individuals to experiment with multimedia creation.

As the internet gained prominence, multimedia found a natural home in the digital realm. Websites, once limited to static text and images, evolved into dynamic, interactive platforms featuring videos, animations, and audio. The rise of streaming services further expanded multimedia's reach, making it possible to deliver rich content to global audiences instantaneously. Platforms like YouTube, Spotify, and Netflix became cultural phenomena, demonstrating the power of multimedia to inform, entertain, and connect people across boundaries. At the same time, social media platforms transformed everyday individuals into multimedia creators, enabling them to share their stories, opinions, and creativity with a global audience.

The modern era of multimedia is defined by its interactivity, personalization, and ubiquity. Advanced technologies such as virtual reality (VR), augmented reality (AR), and artificial intelligence (AI) have pushed the boundaries of what multimedia can achieve. VR immerses users in fully digital environments, enabling them to experience stories and simulations in ways that feel lifelike and participatory. AR overlays digital elements onto the physical world, enhancing real-world experiences with interactive visuals and audio. AI, meanwhile, drives personalization in multimedia by analyzing user preferences and delivering tailored content, whether it's movie recommendations on Netflix or targeted advertisements on social media.

In this digital age, multimedia has become a cornerstone of industries ranging from education and marketing to healthcare and entertainment. In education, multimedia enhances learning by combining visual, auditory, and interactive elements to engage students and cater to diverse learning styles. Educational platforms like Khan Academy and Coursera leverage multimedia to deliver lectures, tutorials, and interactive exercises, making learning more accessible and effective. In marketing, multimedia campaigns combine videos, animations, and social media to create compelling brand narratives and engage consumers. In healthcare, multimedia technologies like 3D modeling and virtual simulations are used for medical training, patient education, and therapy.

A critical aspect of multimedia is its reliance on its core components: digital audio, videography, photography, and text. Digital audio plays a pivotal role in shaping the emotional tone and atmosphere of multimedia content. From background scores in films to voiceovers in educational videos, audio elements add depth and dimension to visual media. Sound design and mixing are essential processes that involve creating, editing, and balancing audio to enhance the overall impact of a multimedia project. Whether it's the immersive sound effects of a video game or the soothing narration of a documentary, digital audio brings multimedia to life.

Videography and photography are equally vital to multimedia, providing the visual foundation for storytelling and communication. Digital videography involves the capture and editing of moving images, whether for films, advertisements, or social media content. Advances in camera technology and editing software have made it easier than ever to produce high-quality video content, empowering both professionals and amateurs to explore their creativity. Photography, on the other hand, captures moments in time, serving as both a standalone art form and a complement to other media types.

In multimedia projects, images and videos work together to convey narratives, evoke emotions, and engage audiences.

Text remains a fundamental component of multimedia, serving as the backbone of communication and the medium through which ideas are structured and expressed. Digital text writing encompasses a wide range of applications, from scripting for videos and animations to creating captions, subtitles, and interactive content. Effective text writing in multimedia requires a careful balance of clarity, brevity, and creativity, ensuring that written content complements and enhances other media forms. The evolution of multimedia is not just a story of technological progress but also a reflection of changing cultural and societal dynamics. As technology has advanced, so too have the expectations and behaviors of audiences. Today's audiences demand content that is not only engaging and informative but also interactive, personalized, and accessible. This shift has driven innovation in multimedia production, with creators constantly exploring new ways to capture attention and deliver value.

However, the rise of multimedia also brings challenges. The sheer volume of multimedia content available today makes it difficult for creators to stand out and for audiences to navigate. Issues such as copyright infringement, digital piracy, and the environmental impact of data storage and streaming are growing concerns. Moreover, the digital divide remains a barrier, preventing many people from accessing the benefits of multimedia technologies due to lack of resources or connectivity.

Despite these challenges, the future of multimedia holds immense promise. Emerging technologies such as 5G networks, blockchain, and advanced AI are poised to further revolutionize how multimedia is created, distributed, and consumed. 5G will enable faster and more reliable streaming of high-quality multimedia content, while blockchain can provide solutions for intellectual property protection and fair compensation for creators. AI will continue to enhance multimedia experiences through innovations in personalization, content generation, and interactivity.

Multimedia is more than a technological innovation—it is a powerful tool for human expression, connection, and understanding. Its evolution reflects humanity's innate desire to communicate ideas, share experiences, and create meaningful connections. As multimedia continues to evolve, it will undoubtedly play a central role in shaping the future of communication, education, and entertainment, driving innovation and enriching lives in ways we have yet to imagine. This unit explores these dynamics in detail, examining the core components of multimedia and their applications, while tracing the journey of this ever-evolving field from its conceptual roots to its current state as a cornerstone of the digital age.

A. Digital Audio: Sound Design and Mixing

Digital audio has become an essential component of multimedia, enabling creators to craft immersive and impactful experiences that resonate deeply with audiences. From movies and video games to podcasts and virtual reality, sound is a powerful medium that shapes how people perceive and engage with content. Digital audio refers to sound that is captured, processed, and reproduced using digital technology, offering greater flexibility, precision, and creativity compared to analog systems.

Sound design and **mixing** are fundamental aspects of digital audio production. Sound design involves creating, manipulating, and integrating audio elements to enhance the emotional and narrative dimensions of a project. Mixing, on the other hand, focuses on blending multiple audio tracks into a cohesive whole, ensuring balance, clarity, and aesthetic appeal. Together, these processes form the backbone of digital audio production, transforming raw sound into polished, professional output.

Digital Audio

Digital audio is a cornerstone of modern multimedia, driving innovation in communication, entertainment, and education. It refers to the sound that has been digitized—converted from analog waveforms into digital signals—so that it can be processed, stored, and transmitted using computers and digital devices. This fundamental transformation allows for unprecedented levels of precision, flexibility, and accessibility, enabling sound to reach global audiences with minimal degradation or loss of quality. The journey of digital audio from its inception to its current form is a testament to technological progress and human ingenuity. The shift from analog to digital technology began in the mid-20th century, marking a pivotal moment in sound recording and reproduction. The digital audio revolution introduced new possibilities, allowing creators to manipulate and enhance sound with greater ease and precision while making it accessible to users across various platforms and devices. Digital audio has now become an integral part of our daily lives, found in everything from music streaming services and podcasts to voice assistants and virtual reality. Its versatility and adaptability have transformed how we interact with sound, making it a critical component of modern multimedia systems.

The Evolution of Digital Audio

The development of digital audio began with the invention of the pulse-code modulation (PCM) system in the 1930s, a technology that remains foundational to digital sound. Initially, digital audio was used primarily in telecommunication to improve the clarity of voice signals transmitted over long distances. It wasn't until the 1970s that digital recording systems became feasible for music and entertainment purposes, with pioneers like Sony and Philips leading the charge.

In 1982, the introduction of the compact disc (CD) revolutionized the music industry, offering higher sound fidelity and durability compared to vinyl records and cassette tapes. This was followed by the advent of the MP3 format in the 1990s, which compressed audio files to make them smaller and more portable, enabling the rise of digital music players like the iPod. Today, digital audio is ubiquitous, supported by advanced codecs, powerful editing software, and versatile playback devices. Streaming platforms like Spotify, Apple Music, and YouTube have further democratized access to digital audio, making it easier than ever to consume, create, and share sound.

Key Components of Digital Audio

Digital audio is defined by several technical components, each of which contributes to its quality, functionality, and applicability. A deep understanding of these components is essential for creators and engineers who wish to optimize audio for specific purposes.

1. Sampling Rate

Sampling rate refers to the number of times per second that an analog sound wave is measured during the digitization process. It is measured in hertz (Hz) or kilohertz (kHz). The higher the sampling rate, the more accurately the digital recording can capture the nuances of the original sound wave.

- **How It Works:**
  - During digitization, continuous analog sound waves are divided into discrete intervals, and the amplitude of the wave is recorded at each interval. These measurements are then converted into numerical data.
  - A high sampling rate ensures that even subtle variations in sound frequency are captured, making the digital representation closer to the original.
- **Common Sampling Rates:**
  - **44.1 kHz:** Standard for CD-quality audio, capable of reproducing frequencies up to 22.05 kHz, which covers the entire range of human hearing.
  - **48 kHz:** Widely used in video production, ensuring compatibility with film and television standards.
  - **96 kHz and above:** Preferred for professional music production and high-resolution audio, offering exceptional detail.
- **Practical Example:**
  - In a live concert recording, a 96 kHz sampling rate captures intricate details like the resonance of string instruments and the subtleties of audience applause, ensuring an immersive listening experience.
- **Limitations:**
  - Higher sampling rates result in larger file sizes, which may not be necessary for casual listening.

2. Bit Depth

Bit depth determines the resolution of each audio sample, affecting the dynamic range and level of detail in the recording. It specifies the number of bits used to represent the amplitude of each sample, with higher bit depths offering more precise representation.

- **How It Works:**
  - Bit depth defines the number of possible amplitude values that can be assigned to each sample. For example, a 16-bit system allows for 65,536 levels, while a 24-bit system provides over 16 million levels.
- **Common Bit Depths:**
  - **16-bit:** Standard for CDs and many consumer audio formats, offering a dynamic range of about 96 dB.
  - **24-bit:** Common in professional recording studios, providing a dynamic range of 144 dB.
  - **32-bit float:** Advanced formats for editing and processing, allowing for exceptionally precise adjustments.

- **Practical Example**:
    - In a film score recording, 24-bit depth ensures that quiet passages and powerful crescendos are captured with equal clarity and without distortion.
- **Challenges**:
    - Higher bit depths require more storage and processing power, which can be a consideration for mobile and low-cost devices.

3. Channels

Channels determine the spatial characteristics of digital audio, ranging from simple mono tracks to complex surround sound systems.

- **Types of Channels**:
    - **Mono**: Single-channel audio where all sounds are combined into one track.
    - **Stereo**: Two-channel audio with distinct left and right tracks, creating a sense of spatial separation.
    - **Surround Sound**: Multi-channel audio (e.g., 5.1 or 7.1 systems) designed to immerse listeners by distributing sound across multiple speakers.
- **Applications**:
    - Mono: Common in telephone systems and voice recordings.
    - Stereo: Used in music playback and general multimedia.
    - Surround Sound: Found in home theaters, gaming setups, and cinema environments.
- **Practical Example**:
    - In a video game, surround sound allows players to hear approaching enemies or environmental cues, enhancing immersion and gameplay strategy.

4. Compression

Compression reduces the file size of digital audio recordings without significantly affecting perceived quality. This process is critical for making audio files easier to store and transmit, particularly in the age of streaming and mobile devices.

- **Types of Compression**:
    - **Lossy Compression**: Discards some audio data to achieve smaller file sizes. Formats include MP3, AAC, and OGG.
    - **Lossless Compression**: Preserves all original audio data while reducing file size. Formats include FLAC and ALAC.
- **Practical Example**:
    - A 3-minute song in uncompressed WAV format might take up 30 MB of storage, while a lossy MP3 version could occupy just 3 MB, making it ideal for portable devices.

- **Limitations:**
    - Lossy compression may result in audible artifacts, particularly at lower bit rates.

Applications of Digital Audio

Digital audio has revolutionized various industries, becoming an indispensable tool for creativity, communication, and innovation. Its applications span numerous fields, each leveraging its unique capabilities to enhance user experiences.

1. **Music Production:**
    - Digital audio workstations (DAWs) like Ableton Live and Pro Tools enable artists to compose, record, and edit music with unparalleled precision.
2. **Film and Television:**
    - Soundtracks, dialogue, and effects are digitally mastered to provide immersive cinematic experiences.
3. **Gaming:**
    - Dynamic audio systems respond to player actions, creating interactive soundscapes.
4. **Education:**
    - Digital audio enhances e-learning platforms with podcasts, audiobooks, and interactive lessons.
5. **Healthcare:**
    - Audio technologies assist in therapies, diagnostics, and accessibility solutions like hearing aids.
6. **Virtual Reality:**
    - Spatial audio in VR environments provides lifelike experiences.

Emerging Trends in Digital Audio

The field of digital audio continues to evolve, driven by advancements in technology:

1. **AI-Powered Tools:**
    - Artificial intelligence is being used for automated mixing, mastering, and even composing music.
2. **Spatial Audio:**
    - 3D sound technologies are gaining popularity in AR/VR and gaming.
3. **Blockchain and NFTs:**
    - Decentralized technologies are being explored for intellectual property management in music and audio content.

Digital audio is more than a technological innovation—it is a medium that empowers human expression and connects people across cultures and boundaries. Its evolution, from early recording systems to cutting-edge AI-driven tools, reflects a journey of continuous innovation and adaptation.

The Role of Sound Design

Sound design is an essential discipline in the realm of multimedia, intertwining artistic creativity and technical mastery to craft auditory elements that elevate the storytelling experience. Sound has the profound ability to evoke emotions, establish atmospheres, and guide the audience's attention, often complementing or even surpassing visual elements in impact. A carefully constructed soundscape can transport listeners to a new world, immerse them in a gripping narrative, or subtly underscore the nuances of a scene. By shaping how audiences perceive and interact with a multimedia project, sound design plays a pivotal role in the overall effectiveness of any creative endeavor. Sound design is not merely about adding sound to visuals; it is about crafting an auditory narrative that interacts harmoniously with visual and textual components. This multidisciplinary practice requires a deep understanding of audio engineering, acoustics, psychology, and storytelling. Sound designers employ cutting-edge tools, software, and techniques to create, manipulate, and integrate audio elements into a cohesive sonic environment.

Key Elements of Sound Design

Sound design encompasses several core components, each serving a specific purpose in enhancing the auditory and emotional experience of multimedia content. These elements, when skillfully combined, create an immersive and compelling soundscape.

1. Dialogue

Dialogue serves as the verbal backbone of many multimedia projects, providing critical context, advancing the narrative, and conveying emotional depth. Clarity, authenticity, and synchronization are key to delivering effective dialogue that engages audiences.

- **Role in Sound Design**:
    - Ensures that spoken words are audible and free from background noise or distortions.
    - Uses equalization (EQ) to enhance vocal clarity and timbre, ensuring the dialogue cuts through other sound layers.
- **Applications**:
    - Films and television rely heavily on clear and emotionally resonant dialogue to establish character relationships and advance plots.
    - In video games, dialogue often includes branching narratives and player interactions, requiring intricate sound design to maintain immersion.
- **Example**:
    - In animated films, voice acting is paired with sound design to bring characters to life, such as the witty exchanges in Pixar's *Toy Story*.

2. Sound Effects (SFX)

Sound effects add layers of realism, excitement, or abstraction to a scene, enriching the audience's sensory experience. From subtle environmental cues to dramatic action sequences, SFX are critical for creating a believable and engaging world.

- **Role in Sound Design**:
    - Augments the visual narrative by simulating real-world sounds or inventing unique auditory cues for fictional scenarios.
    - Employs techniques such as layering, pitch shifting, and time stretching to create distinctive effects.
- **Applications**:
    - Action films use sound effects to intensify explosions, car chases, and fight sequences.
    - Horror games leverage unsettling sounds to heighten tension and fear.
- **Example**:
    - The lightsaber hum in *Star Wars* is a classic example of iconic SFX that enhances the sci-fi aesthetic of the series.

3. Ambient Sounds

Ambient sounds, often subtle and understated, establish a sense of place and mood, immersing the audience in the environment. These background noises provide context without demanding attention, subtly enhancing the realism of a scene.

- **Role in Sound Design**:
    - Fills sonic gaps to prevent unnatural silence, maintaining the continuity and flow of a scene.
    - Helps anchor the audience in a specific location or setting through auditory cues.
- **Applications**:
    - Documentaries use ambient sounds to reflect the natural environments being depicted.
    - Video games employ dynamic ambient audio to adapt to the player's actions and surroundings.
- **Example**:
    - In a nature documentary, the sound of waves crashing on the shore complements visuals of a serene beach.

4. Foley

Foley is the process of recreating everyday sounds in post-production to enhance the realism and authenticity of a project. Named after sound-effects pioneer Jack Foley, this art form relies on ingenuity and precision to synchronize audio with visuals seamlessly.

- **Role in Sound Design**:
  - Replaces or supplement sounds that are impractical to record during filming.
  - Ensures that minor details, such as footsteps or rustling fabric, contribute to the narrative's believability.
- **Applications**:
  - Films and television shows use Foley artists to create sounds that blend naturally with on-screen actions.
  - Animated movies rely entirely on Foley to produce sound effects for character movements and interactions.
- **Example**:
  - In *Jurassic Park*, Foley artists used animal recordings and Foley techniques to create realistic dinosaur sounds.

5. Music

Music is a powerful element of sound design, capable of conveying emotions, building tension, and emphasizing key moments. Whether through original compositions or licensed tracks, music adds depth and dimension to multimedia projects.

- **Role in Sound Design**:
  - Sets the emotional tone, guiding the audience's reactions to the narrative.
  - Uses tempo, key, and dynamics to align with the pacing and mood of the visuals.
- **Applications**:
  - Movie scores enhance storytelling, with leitmotifs representing characters or themes.
  - Commercials use upbeat tracks to grab attention and reinforce brand messaging.
- **Example**:
  - The foreboding music in *Jaws* signals the impending danger of the shark, creating suspense and fear.

6. Synthesized Sounds

Synthesized sounds are created using electronic instruments or software, enabling designers to produce auditory effects that do not exist in nature. These sounds are often used in genres like science fiction and fantasy to convey futuristic or otherworldly environments.

- **Role in Sound Design**:
  - Provides unique auditory signatures for characters, objects, or locations.
  - Offers endless creative possibilities by manipulating waveforms and effects.

- **Applications**:
  - Sci-fi movies use synthesized sounds to create futuristic weapon effects or alien vocalizations.
  - Music producers incorporate synthesized elements into clectronic dance music (EDM) and experimental genres.

- **Example**:
  - The atmospheric soundscapes in *Blade Runner* rely heavily on synthesized audio to define the film's dystopian aesthetic.

Techniques in Sound Design

Sound design is a meticulous process that employs various techniques to craft and refine audio elements. These techniques enhance the quality, clarity, and emotional impact of the soundscape.

- **Layering**:
  - Combines multiple audio elements to create a rich and complex auditory texture.
  - Example: The sound of a thunderstorm may include layered recordings of rain, thunderclaps, and wind.
- **Equalization (EQ)**:
  - Adjusts the balance of frequencies in a sound to enhance clarity and remove unwanted noise.
  - Example: Boosting low frequencies to emphasize the rumble of an approaching train.
- **Reverb and Delay**:
  - Adds spatial effects to simulate the acoustics of different environments.
  - Example: Using reverb to make a character's voice echo in a large cave.
- **Pitch Shifting**:
  - Alters the pitch of a sound to create dramatic or surreal effects.
  - Example: Lowering the pitch of a roar to make it sound more menacing.
- **Dynamic Range Compression**:
  - Balances the volume of different audio elements to maintain consistency and prevent distortion.

The Impact of Sound Design

Sound design transcends the auditory realm, shaping how audiences experience and interpret multimedia content. Its influence extends to various fields, including:

- **Film and Television**:
  - Sound design enriches visual storytelling, from the subtle ambiance of a quiet scene to the thunderous intensity of a battle sequence.
- **Video Games**:
  - Interactive audio reacts to player actions, enhancing immersion and engagement.
  - Example: In *The Last of Us*, sound design plays a crucial role in building tension and emotional resonance.
- **Virtual Reality (VR)**:
  - Spatial audio creates realistic environments, making users feel as though they are physically present in the digital world.
- **Advertising**:
  - Catchy jingles and sound effects reinforce brand identity and leave lasting impressions on audiences.

Sound design is an indispensable art form that breathes life into multimedia projects. By combining creativity, technical expertise, and a deep understanding of storytelling, sound designers create auditory experiences that captivate, inform, and inspire. As technology continues to advance, the possibilities for innovation in sound design are limitless, promising new dimensions of immersion and emotional impact in multimedia content.

Techniques in Sound Design

Sound design is a nuanced and creative discipline that involves the manipulation and crafting of auditory elements to produce compelling and immersive audio experiences. Techniques used in sound design help transform raw sounds into refined components that align with the creative vision of a multimedia project. By employing these techniques, sound designers can add depth, emotion, and realism to auditory landscapes, making them more impactful and engaging for audiences.

1. Layering

Layering is one of the foundational techniques in sound design, where multiple audio elements are combined to create rich and complex sound textures. By stacking different sounds on top of one another, sound designers can build intricate auditory experiences that resonate with the audience.

- **Process**:
  - Designers select individual sound samples, each contributing a unique quality to the final composite sound.
  - Sounds are arranged in layers, often with adjustments to volume, pitch, and timing to ensure cohesion.
  - Effects like reverb, distortion, or EQ may be applied to individual layers to enhance their contribution to the overall sound.

- **Applications**:
    - **Film and Animation**: Layering is frequently used to create fantastical sounds, such as the roar of a mythical creature or the sound of a futuristic spaceship.
    - **Video Games**: Interactive soundscapes, like those in battle scenes, rely heavily on layered audio to convey chaos and immersion.
- **Example**:
    - The iconic roar of a dragon in fantasy films might be achieved by layering various audio elements, including the growls of animals like lions, the sound of thunder, and recordings of rushing wind. Together, these layers form a powerful, otherworldly sound.

2. Equalization (EQ)

Equalization, or EQ, involves adjusting the balance of frequencies in a sound to enhance clarity, remove unwanted noise, and emphasize specific audio elements. This technique allows sound designers to sculpt sounds so that they fit harmoniously within a larger auditory mix.

- **Process**:
    - Audio frequencies are divided into bands (low, mid, high), and adjustments are made to boost or reduce specific ranges.
    - Problematic frequencies, such as background hums or sharp peaks, are identified and minimized.
- **Applications**:
    - **Music Production**: EQ ensures that vocals and instruments do not clash, creating a balanced and pleasing composition.
    - **Dialogue Clarity**: In films and podcasts, EQ is used to make dialogue clear and prominent, even when layered with music and sound effects.
- **Example**:
    - In a scene depicting an earthquake, the rumbling sound is made more intense by boosting low-frequency ranges, giving the audience a visceral sense of the tremor's power.

3. Reverb and Echo

Reverb and echo are spatial effects used to simulate different acoustic environments, adding depth and realism to audio. While reverb creates a sense of space by mimicking the way sound reflects off surfaces, echo replicates distinct repetitions of a sound over time.

- **Process**:
    - Designers apply reverb to a sound to mimic the acoustics of specific environments, such as a small room, a concert hall, or an open field.

- Echo effects are added by delaying repetitions of the original sound, creating a distinct trailing effect.

- **Applications**:
  - **Environmental Soundscapes**: Reverb and echo are essential for establishing location-specific atmospheres in films, games, and VR experiences.
  - **Voice Effects**: Reverb can make a voice sound ethereal, while echo can suggest distance or eeriness.
- **Example**:
  - A scene set in a large cathedral might use reverb to make characters' voices sound as if they are reverberating through the vast, hollow space, enhancing the grandeur of the setting.

4. Pitch Shifting

Pitch shifting alters the pitch of a sound to make it higher or lower, enabling designers to create dramatic, surreal, or entirely new auditory effects. This technique is commonly used to modify voices, music, or sound effects.

- **Process**:
  - Pitch shifting software or plugins are used to adjust the frequency of a sound wave without changing its tempo (time stretching) or in conjunction with tempo alterations.
  - Pitch can be shifted up to make a sound brighter or more energetic, or shifted down to make it deeper and more menacing.
- **Applications**:
  - **Voice Modulation**: In sci-fi and fantasy genres, pitch shifting is used to create alien voices or monstrous growls.
  - **Music Effects**: Altering the pitch of instruments can add variety and dynamics to compositions.
- **Example**:
  - In a horror movie, a child's laughter might be pitch-shifted upward to create a creepy, unsettling effect, contrasting the innocence of the sound with the darker context of the scene.

5. Dynamic Range Compression

Dynamic range compression is the process of reducing the difference between the loudest and softest parts of an audio track. This technique ensures consistency in volume levels, preventing sounds from being too overpowering or too faint to hear.

- **Process**:
  - A compressor is used to identify and attenuate peaks in volume while maintaining or boosting quieter parts of the track.

- Settings such as threshold, ratio, attack, and release are adjusted to achieve the desired effect.

- **Applications**:
  - **Film and Television**: Compression ensures that dialogue remains audible over background music or action sequences.
  - **Broadcasting**: Audio for radio and television is compressed to maintain consistent volume levels across different programs.
- **Example**:
  - In a movie, dynamic range compression might be applied to dialogue during an intense action scene to ensure the characters' voices remain clear amid explosions and gunfire.

Advanced Techniques and Tools

In addition to the core techniques, sound designers often employ advanced tools and methods to enhance their work:

- **Automation**: Dynamically adjusts parameters like volume, panning, and effects over time to match the visual flow of a project.
- **Sound Morphing**: Combines elements of two or more sounds to create entirely new auditory effects, often used in sci-fi or fantasy settings.
- **Spatial Audio**: Creates 3D soundscapes for immersive experiences in virtual reality (VR) and augmented reality (AR).

Impact of Sound Design Techniques

Each of these techniques contributes to the overall effectiveness of a multimedia project, enhancing its emotional resonance, realism, and narrative depth. When used skillfully, sound design techniques ensure that audio not only supports but elevates the visual and textual elements of a project, creating a cohesive and engaging experience for audiences. By mastering these techniques, sound designers can craft auditory environments that captivate listeners, transport them to new worlds, and leave lasting impressions. As technology continues to evolve, the possibilities for innovation in sound design will only expand, promising even more dynamic and immersive auditory experiences in the future.

The Role of Mixing

Mixing is a cornerstone of audio production that brings together multiple audio elements into a harmonious and polished soundscape. This process ensures that all components—whether dialogue, music, sound effects, or ambient sounds—are well-balanced, complementary, and aligned with the creative intent of the project. Mixing transforms raw, unprocessed audio into a cohesive auditory experience that captivates audiences and enhances the overall impact of multimedia content. Effective mixing is not just about technical precision; it is also an artistic endeavor. The mixer must understand the emotional and narrative goals of the project, ensuring that the audio not only sounds good but also serves the story. Whether in film, music, video games, or podcasts, a well-mixed audio track can guide the listener's attention, evoke emotions, and create an immersive environment.

Steps in the Mixing Process

The mixing process involves several stages, each contributing to the clarity, balance, and aesthetic quality of the final audio product. These steps are typically performed in a digital audio workstation (DAW) using specialized tools and plugins.

1. Organization

The first step in mixing is organizing the audio tracks to streamline the workflow. A well-structured session ensures efficiency and minimizes errors during the mixing process.

- **What It Involves:**
    - Labeling tracks clearly (e.g., "Dialogue," "Background Music," "Ambience").
    - Grouping related tracks into folders or buses for easier management.
    - Eliminating unwanted noise or redundant audio clips.
- **Tools:**
    - DAWs like Pro Tools, Logic Pro, and Adobe Audition allow users to color-code tracks, set markers, and create templates for consistent organization.
- **Importance:**
    - Proper organization saves time and allows the mixer to focus on creative decisions rather than logistical challenges.

2. Balancing Levels

Balancing levels involves adjusting the volume of individual tracks to ensure that no single element dominates the mix unless intended. This step is crucial for creating a natural and cohesive soundscape.

- **What It Involves:**
    - Ensuring that dialogue is clear and intelligible, especially in multimedia projects like films or podcasts.
    - Adjusting the volume of background music and sound effects to complement rather than overpower the primary elements.
- **Example:**
    - In a film scene, dialogue often takes precedence over background music or ambient sounds to maintain clarity and focus.
- **Tools:**
    - Volume faders and gain controls in DAWs allow precise adjustments.
    - Meters and spectrum analyzers help monitor levels and avoid clipping or distortion.

3. Panning

Panning refers to the placement of audio within the stereo or surround sound field, creating a sense of direction and space. This technique adds depth and realism to the audio, making it more engaging for listeners.

- **What It Involves**:
  - Distributing sounds across the left and right channels in a stereo mix.
  - Positioning sounds in a surround sound setup to match on-screen action or create immersive environments.
- **Example**:
  - The sound of a car passing from left to right in a film can be panned accordingly to reflect its movement in the scene.
- **Applications**:
  - **Film**: Surround sound mixes use panning to place sounds like explosions or footsteps within a 3D auditory space.
  - **Music**: Instruments are panned to create a balanced stereo image, such as placing guitars slightly to the left and keyboards to the right.

4. EQ and Filtering

Equalization (EQ) and filtering are used to enhance the tonal quality of audio tracks by adjusting their frequency content. This step ensures that each element in the mix occupies its own sonic space, preventing muddiness or clutter.

- **What It Involves**:
  - Boosting or cutting specific frequency ranges to enhance clarity and remove unwanted noise.
  - Using high-pass or low-pass filters to isolate desired frequencies.
- **Example**:
  - In a podcast, reducing low-frequency rumble (e.g., from air conditioning) makes dialogue sound cleaner and more professional.
- **Tools**:
  - Parametric EQ plugins like FabFilter Pro-Q or stock EQ tools in DAWs allow precise frequency adjustments.

5. Effects Processing

Effects processing involves adding creative or corrective effects to audio tracks to enhance their character and emotional impact. Effects can range from subtle reverb to dramatic modulation, depending on the project's needs.

- **Common Effects**:
    - **Reverb**: Simulates the acoustics of a specific environment, such as a concert hall or cave.
    - **Delay**: Creates echoes that can add depth or rhythmic interest.
    - **Chorus**: Adds richness by duplicating and slightly detuning the original sound.
- **Example**:
    - In a music track, reverb on vocals can create a sense of space and emotion, while delay can add rhythmic complexity to a guitar solo.

6. Automation

Automation allows dynamic adjustments to parameters like volume, panning, and effects over time. This step ensures that the mix evolves with the narrative or musical structure, maintaining interest and coherence.

- **What It Involves**:
    - Automating volume fades during transitions between scenes or songs.
    - Modulating effects like reverb or EQ to reflect changes in mood or setting.
- **Example**:
    - In a film, the background music might fade out gradually as dialogue becomes more prominent during an emotional moment.
- **Tools**:
    - DAWs provide automation lanes where users can draw or record parameter changes in real time.

7. Mastering

Mastering is the final stage of audio production, preparing the mixed audio for distribution across various platforms and playback systems. This step ensures consistency in loudness, tonal balance, and format compatibility.

- **What It Involves**:
    - Applying compression and limiting to achieve optimal loudness without distortion.
    - Equalizing the overall mix to ensure it translates well across different devices (e.g., headphones, speakers, or cinemas).
    - Encoding the audio in the appropriate format, such as MP3, WAV, or AAC.
- **Tools**:
    - Mastering plugins like iZotope Ozone and Waves L3 Ultramaximizer are widely used for professional mastering.

- **Importance**:
    - A mastered mix ensures that the audio sounds polished and professional, regardless of where or how it is played.

The Impact of Mixing

Mixing plays a pivotal role in defining the overall quality and impact of multimedia projects. A well-mixed audio track enhances storytelling, reinforces emotions, and maintains audience engagement. Whether it's the subtle ambiance in a dramatic scene or the pulsating beats of a dance track, the mix determines how effectively the audio resonates with listeners. In film and television, mixing ensures that dialogue, music, and sound effects work together seamlessly to support the narrative. In music production, it brings out the best in each instrument and vocal performance. In video games, mixing creates dynamic soundscapes that adapt to player actions, enhancing immersion. As a blend of technical skill and artistic intuition, mixing is an indispensable part of audio production that shapes the way audiences experience and connect with multimedia content.

B. Digital Videography and Photography

Digital videography and photography are integral components of multimedia, serving as powerful tools for storytelling, documentation, and artistic expression. These disciplines have been revolutionized by advancements in digital technology, offering unparalleled flexibility, precision, and creative possibilities. Digital videography involves capturing and editing moving images, while digital photography focuses on still images. Together, they form the visual backbone of modern communication, spanning industries such as film, advertising, education, journalism, and social media. The transition from analog to digital methods has redefined how visuals are captured, processed, and shared. Today, digital cameras, editing software, and online platforms empower creators to produce high-quality content accessible to global audiences. Whether it's a breathtaking photograph or a compelling video, digital videography, and photography enable creators to craft visual narratives that resonate deeply with viewers.

Digital Videography

Digital videography is a cornerstone of modern multimedia, encompassing the capture, editing, and production of video content using digital technology. Unlike traditional analog or film-based methods, which rely on chemical processes and physical storage mediums, digital videography utilizes electronic sensors to convert light into digital data. This transition to digital formats has transformed video production, offering unparalleled efficiency, flexibility, and creative control. As a result, digital videography has become accessible to professionals and hobbyists alike, empowering them to produce content that ranges from cinematic masterpieces to social media clips. The evolution of digital videography has aligned with advancements in camera technology, editing software, and distribution platforms. Today, videos dominate the digital landscape, serving as powerful tools for storytelling, education, marketing, and entertainment.

Key Components of Digital Videography

Digital videography involves a blend of hardware, software, and artistic vision. Understanding its core components is essential for mastering the craft and achieving professional-quality results.

1. Digital Cameras

Digital cameras are the primary tools for capturing video in digital videography. They use image sensors to convert light into digital signals, which are then processed and stored as video files. Two common types of sensors used in digital cameras are **CMOS (Complementary Metal-Oxide Semiconductor)** and **CCD (Charge-Coupled Device)**. CMOS sensors are widely used due to their efficiency and ability to support higher resolutions.

- **Types of Digital Cameras**:
  - **DSLR Cameras**: Known for their versatility, these cameras offer interchangeable lenses, high-quality sensors, and manual controls.
    - Example: Canon EOS 5D Mark IV, a favorite among filmmakers and photographers.
  - **Mirrorless Cameras**: Compact and lightweight, these cameras provide similar features to DSLRs with advanced autofocus and video capabilities.
    - Example: Sony Alpha series, popular for its 4K video recording.
  - **Cinema Cameras**: Designed for professional film production, these cameras deliver exceptional image quality and dynamic range.
    - Example: RED cameras and ARRI Alexa, used in Hollywood blockbusters.
- **Importance**:
  - The choice of camera determines the resolution, frame rate, and overall quality of the footage, making it a critical decision for videographers.

2. Resolution

Resolution defines the clarity and detail of a video by specifying the number of pixels in each frame. Higher resolutions offer sharper and more detailed images, making them ideal for high-definition content.

- **Common Formats**:
  - **HD (1920x1080)**: Standard resolution for most online videos and television broadcasts.
  - **4K (3840x2160)**: Provides four times the resolution of HD, delivering exceptional detail and clarity. Widely used in cinema and professional projects.
  - **8K (7680x4320)**: The cutting-edge resolution for ultra-high-definition content, suitable for large screens and immersive experiences.
- **Applications**:
  - HD is sufficient for most social media platforms and casual projects.
  - 4K and 8K are preferred for cinematic productions, advertising, and virtual reality (VR) applications.

- **Importance**:
    - The resolution impacts not only the visual quality but also the file size and processing requirements of the video.

3. Frame Rate

Frame rate, measured in frames per second (fps), determines the smoothness and cinematic quality of a video. It refers to the number of individual frames captured or displayed per second.

- **Common Frame Rates**:
    - **24 fps**: Standard for films, offering a cinematic look with slight motion blur.
    - **30 fps**: Common in television and online videos, providing a sharper appearance.
    - **60 fps and higher**: Used for slow-motion effects, sports broadcasts, and gaming.
- **Creative Considerations**:
    - Higher frame rates capture more detail in motion, making them ideal for action scenes.
    - Lower frame rates convey a traditional cinematic feel, enhancing the storytelling experience.
- **Example**:
    - A music video shot at 60 fps can include slow-motion sequences for dramatic effect, while a narrative film might stick to 24 fps for a classic look.

4. Lenses

The lens is a critical component of a digital camera, influencing the perspective, focus, and depth of field in a video. The choice of lens significantly impacts the aesthetics and storytelling of a project.

- **Types of Lenses**:
    - **Wide-Angle Lenses**: Capture expansive views, making them ideal for landscapes and establishing shots.
    - **Telephoto Lenses**: Zoom in on distant subjects, often used for close-ups and sports.
    - **Prime Lenses**: Fixed focal length lenses known for their sharpness and superior image quality.
- **Factors to Consider**:
    - **Aperture**: Determines how much light enters the lens, affecting the brightness and depth of field.
    - **Focal Length**: Influences the field of view and perspective.
- **Example**:
    - A wide-angle lens is perfect for capturing a cityscape, while a telephoto lens excels in isolating a subject from the background.

5. Lighting

Lighting is a vital element in videography, shaping the mood, clarity, and overall visual appeal of a video. Proper lighting can transform an ordinary scene into a cinematic masterpiece.

- **Types of Lighting**:
  - **Natural Light**: Outdoor shoots often rely on sunlight, which can be diffused or enhanced with reflectors.
  - **Artificial Light**: Studio setups use tools like softboxes, LED panels, and spotlights for precise control.
- **Lighting Techniques**:
  - **Three-Point Lighting**: Involves key light, fill light, and backlight to create depth and dimension.
  - **High-Key Lighting**: Bright and even, often used in comedies and commercials.
  - **Low-Key Lighting**: Creates dramatic shadows, commonly seen in thrillers and noir films.
- **Importance**:
  - The quality and direction of light affect the visibility, mood, and emotional impact of a scene.

6. Editing Software

Editing software is an indispensable tool in digital videography, allowing creators to refine raw footage into polished videos. Post-production involves cutting, color grading, adding effects, and synchronizing audio.

- **Popular Tools**:
  - **Adobe Premiere Pro**: Known for its versatility and professional features.
  - **Final Cut Pro**: Apple's editing software, favored for its speed and user-friendly interface.
  - **DaVinci Resolve**: Renowned for its advanced color grading capabilities.
- **Key Features**:
  - Timeline editing for precise cuts and transitions.
  - Color correction and grading to achieve the desired visual tone.
  - Integration of visual effects, graphics, and animations.
- **Example**:
  - A travel vlog might use Premiere Pro to blend drone footage, time-lapse sequences, and music into an engaging narrative.

The Significance of Digital Videography

Digital videography has democratized video production, enabling creators across all skill levels to share their stories and ideas. Its applications are vast, spanning industries such as:

- **Entertainment**: From blockbuster films to indie projects, digital videography is the backbone of visual storytelling.
- **Education**: Online courses, tutorials, and documentaries rely on high-quality videos to engage learners.
- **Marketing**: Brands use video content to captivate audiences, showcase products, and drive conversions.
- **Social Media**: Platforms like YouTube, TikTok, and Instagram thrive on video content, making videography a powerful tool for creators and influencers.

Applications of Digital Videography

Digital videography has become a versatile and indispensable tool in a variety of fields, leveraging its ability to convey complex ideas, evoke emotions, and captivate audiences through dynamic visual storytelling. Its applications span across industries, shaping how we consume information, communicate ideas, and share experiences in the digital age. From cinema to social media, digital videography plays a pivotal role in defining modern multimedia and its impact on society.

1. Film and Television

Digital videography has revolutionized the film and television industry, providing creators with tools to produce visually stunning content that captivates global audiences. The ability to shoot in high resolution, coupled with advanced post-production techniques, has elevated storytelling to new heights.

- **Uses**:
  - Producing feature films, documentaries, and TV shows with unparalleled visual quality.
  - Capturing intricate scenes using advanced cameras and drones for aerial shots.
- **Advantages**:
  - Cost Efficiency: Digital workflows eliminate the need for film stock and enable faster editing.
  - Flexibility: Instant playback during shoots allows directors to make adjustments on the spot.
  - Accessibility: High-quality digital cameras are more affordable, democratizing filmmaking.
- **Example**:
  - *Avengers: Endgame*, one of the highest-grossing films of all time, was shot using high-resolution digital cameras, creating visually immersive effects that brought the Marvel Cinematic Universe to life.

2. Marketing and Advertising

In the fast-paced world of marketing, video content is a powerful tool for engaging audiences, showcasing products, and building brand loyalty. Digital videography enables businesses to create impactful campaigns that resonate with consumers across various platforms.

- **Uses**:
  - Product demonstrations and launch videos that highlight key features and benefits.
  - Explainer videos that simplify complex concepts for potential customers.
  - Promotional campaigns tailored to specific audiences on platforms like YouTube, Facebook, and Instagram.
- **Advantages**:
  - Enhanced Engagement: Videos capture attention more effectively than static images or text.
  - Global Reach: Online video ads can target audiences worldwide with tailored content.
  - Measurable Impact: Analytics tools track viewer engagement and ROI for video campaigns.
- **Example**:
  - Apple's product launch videos blend sleek videography with storytelling, demonstrating the capabilities of new devices while reinforcing the brand's innovative image.

3. Social Media

Social media platforms are dominated by short, engaging video content, making digital videography a cornerstone of online communication. From user-generated content to professional marketing campaigns, videos drive interaction, virality, and community building.

- **Uses**:
  - Creating attention-grabbing short videos for platforms like TikTok, Instagram Reels, and YouTube Shorts.
  - Sharing behind-the-scenes footage, tutorials, and interviews to connect with followers.
  - Showcasing live events or promotions through real-time video streaming.
- **Advantages**:
  - Broad Audience: Social media videos can reach millions of viewers within minutes.
  - Interactivity: Features like comments, likes, and shares encourage engagement and feedback.
  - Cost-Effective: Even small businesses can produce impactful social media videos with minimal budgets.

- **Example**:
  - Influencers on Instagram use creative videography to craft visually stunning Reels, boosting their engagement and expanding their audience reach.

4. Education and Training

E-learning has grown exponentially, relying on digital videography to deliver interactive, engaging, and accessible content for learners worldwide. Videos help bridge the gap between traditional classroom instruction and digital education.

- **Uses**:
  - Creating tutorials, lectures, and instructional videos for online courses and workshops.
  - Recording simulations and demonstrations for technical and vocational training.
  - Enhancing corporate training programs with engaging multimedia content.
- **Advantages**:
  - Accessibility: Learners can access educational videos anytime, anywhere.
  - Versatility: Videos cater to various learning styles, incorporating visuals, text, and audio.
  - Retention: Studies show that learners retain more information through video-based instruction.

- **Example**:
  - Platforms like Coursera and Khan Academy use high-quality videography to create engaging course content that caters to global learners.

5. Journalism

Digital videography has transformed journalism, enabling real-time reporting and immersive storytelling that captures the attention of audiences. Video content enhances the credibility and impact of news stories by bringing them to life with visual evidence and emotional resonance.

- **Uses**:
  - Recording breaking news events and broadcasting them in real-time.
  - Producing in-depth investigative reports with interviews, archival footage, and visual analysis.
  - Documenting social issues and cultural stories for digital platforms.
- **Advantages**:
  - Speed: Digital tools allow journalists to record, edit, and publish stories within hours.
  - Engagement: Videos evoke empathy and understanding, making complex issues relatable.

- Reach: Online videos can be shared globally, amplifying the impact of a story.

- **Example**:
  - News organizations like BBC and Al Jazeera use digital videography to report on global events, providing viewers with comprehensive visual narratives.

Emerging Applications and Future Trends

Digital videography is constantly evolving, with emerging applications expanding its influence across industries:

1. **Healthcare**:
   - Medical professionals use video content for training, patient education, and telemedicine consultations.
   - Example: Surgical procedures recorded with advanced videography techniques for instructional purposes.
2. **Real Estate**:
   - Virtual tours created with digital videography help buyers explore properties remotely.
   - Example: Real estate agencies use drone videography to showcase properties from unique perspectives.
3. **Event Coverage**:
   - Weddings, conferences, and live performances are captured with cinematic videography, preserving memories and experiences.
   - Example: Highlight reels from events are shared online, reaching broader audiences.
4. **Gaming and Esports**:
   - Videography enhances the promotion and broadcast of gaming events, tournaments, and live streams.
   - Example: Esports tournaments use dynamic video production to create excitement and engage viewers.
5. **Virtual and Augmented Reality**:
   - VR and AR applications rely on immersive videography to create realistic and interactive environments.
   - Example: Travel experiences use 360-degree videos to offer virtual tours of exotic locations.

Digital Photography

Digital photography is the process of capturing still images using digital cameras, where light is recorded as electronic data instead of undergoing chemical reactions on film. This technological advancement has transformed photography, making it accessible to professionals and amateurs alike. With digital photography, users can instantly review, edit, and share their work, offering unprecedented flexibility and creative possibilities. Unlike traditional photography, which requires developing film in darkrooms, digital photography provides immediate results. The integration of advanced hardware like high-resolution sensors and lenses, combined with powerful post-processing tools, allows photographers to produce stunning images tailored to various applications. Digital photography spans multiple domains, from art and journalism to marketing and science, demonstrating its versatility and significance.

Key Components of Digital Photography

Digital photography relies on a combination of hardware, software, and creative techniques to produce high-quality images. Understanding these components is essential for achieving professional results.

1. Digital Cameras

Digital cameras are the foundation of modern photography, offering a range of features and functionalities that cater to different skill levels and use cases.

- **Types of Digital Cameras**:
    - **Point-and-Shoot Cameras**: Compact and user-friendly, these cameras are ideal for casual photographers.
        - Example: Canon PowerShot series.
    - **DSLR Cameras**: Known for their versatility and high-quality output, DSLRs offer interchangeable lenses, manual controls, and superior performance.
        - Example: Nikon D850.
    - **Mirrorless Cameras**: Lightweight alternatives to DSLRs with comparable features and advanced autofocus systems.
        - Example: Sony Alpha series.
    - **Smartphone Cameras**: Equipped with multiple lenses and AI enhancements, smartphones have become powerful tools for everyday photography.
        - Example: Apple iPhone Pro series.
- **Key Features**:
    - Autofocus ensures sharp images, even in challenging conditions.
    - Burst mode captures multiple frames per second, ideal for action shots.
    - Image stabilization minimizes blurriness caused by camera shake.

2. Resolution

Resolution, measured in megapixels (MP), determines the level of detail and clarity in a photograph. Higher resolutions produce sharper images suitable for large prints and detailed editing.

- **Resolution Categories**:
    - **Entry-Level Cameras**: 12–20 MP, sufficient for casual photography and social media.
    - **Professional Cameras**: 40 MP and above, ideal for commercial and fine art photography.
- **Importance of Resolution**:
    - A higher resolution allows for cropping and resizing without significant loss of quality.
    - Lower resolutions are more practical for online sharing due to reduced file sizes.

3. Image Sensors

The image sensor is a crucial component that captures light and converts it into digital signals. The size and type of sensor significantly affect image quality, low-light performance, and depth of field.

- **Types of Sensors**:
    - **Full-Frame Sensors**: Match the size of 35mm film, offering superior image quality and minimal noise.
        - Example: Canon EOS R5.
    - **APS-C Sensors**: Slightly smaller than full-frame sensors, balancing quality and cost.
        - Example: Fujifilm X-T4.
    - **Micro Four Thirds Sensors**: Smaller sensors used in compact systems like Panasonic Lumix, providing portability.
- **Sensor Size Impact**:
    - Larger sensors capture more light, resulting in better low-light performance and richer colors.

4. Lenses

Lenses are pivotal in shaping the visual composition of a photograph, influencing sharpness, perspective, and depth of field.

- **Types of Lenses**:
    - **Macro Lenses**: Designed for extreme close-ups, capturing fine details in subjects like insects or flowers.
    - **Wide-Angle Lenses**: Ideal for landscapes and architectural photography, offering a broader field of view.
    - **Telephoto Lenses**: Zoom in on distant subjects, commonly used in wildlife and sports photography.
    - **Fisheye Lenses**: Create a distorted, spherical effect, often used for artistic or experimental shots.

- **Aperture**:
    - Measured in f-stops, the aperture controls the amount of light entering the lens and affects the depth of field.
    - Example: A wide aperture (e.g., f/1.8) creates a shallow depth of field, isolating the subject against a blurred background.

5. Lighting

Lighting is essential in photography, defining the mood, texture, and visibility of an image. Proper lighting techniques can dramatically enhance the quality and impact of a photograph.

- **Types of Lighting**:
    - **Natural Light**: Sunlight offers a dynamic and versatile source for outdoor photography.
    - **Artificial Light**: Studio setups use ring lights, LED panels, and flash units for controlled environments.
    - **Reflectors**: Redirect light to fill shadows and add balance to a composition.
- **Lighting Techniques**:
    - **Golden Hour Photography**: Captures warm, diffused sunlight during sunrise or sunset.
    - **Backlighting**: Positions the light source behind the subject for silhouettes or dramatic effects.

6. Post-Processing

Post-processing involves enhancing photographs using editing software to correct imperfections, adjust colors, and apply creative effects.

- **Popular Tools**:
    - **Adobe Photoshop**: A versatile platform for detailed edits, such as retouching and compositing.
    - **Adobe Lightroom**: Focuses on batch editing, color correction, and organization.
    - **Capture One**: Preferred by professionals for its advanced color grading capabilities.
- **Editing Techniques**:
    - Adjusting exposure and contrast to improve brightness and shadows.
    - Cropping and straightening to refine composition.
    - Applying filters and effects to achieve a specific artistic style.

Applications of Digital Photography

Digital photography is a versatile medium with applications across various fields, reflecting its artistic, practical, and commercial significance.

1. Portrait Photography

Portrait photography captures the essence of individuals or groups, often highlighting their expressions and personalities.

- **Use Cases**:
  - Professional headshots for resumes, LinkedIn profiles, and corporate branding.
  - Artistic portraits for exhibitions or personal projects.
- **Example**:
  - Photographers use DSLR cameras with wide-aperture lenses to create sharp, well-lit portraits with blurred backgrounds.

2. Product Photography

Product photography showcases items for e-commerce, advertising, and cataloging. High-quality images can significantly influence consumer behavior and sales.

- **Use Cases**:
  - Amazon and Etsy listings rely on clean, detailed images to attract buyers.
  - Lifestyle photography places products in real-world contexts to enhance appeal.
- **Example**:
  - A macro lens captures the intricate details of a wristwatch for an online advertisement.

3. Travel and Landscape Photography

Travel and landscape photography documents natural and urban environments, often for personal expression, tourism campaigns, or editorial features.

- **Use Cases**:
  - Magazines like *National Geographic* showcase breathtaking landscapes.
  - Travel bloggers capture iconic landmarks to inspire wanderlust.
- **Example**:
  - Wide-angle lenses and tripods help photographers capture expansive vistas with clarity and depth.

4. Event Photography

Event photography preserves memories of occasions like weddings, concerts, and corporate gatherings.

- **Use Cases**:
  - Wedding photographers create albums that narrate the day's story.
  - Concert photographers capture the energy and emotion of live performances.

- **Example**:
  - Using a telephoto lens, a photographer captures candid moments from a distance without disturbing the event.

5. Photojournalism

Photojournalism tells stories through impactful images, documenting events, cultures, and human experiences for news outlets and editorial publications.

- **Use Cases**:
  - War photographers highlight the realities of conflict zones.
  - Street photographers capture spontaneous moments of urban life.
- **Example**:
  - A photojournalist uses a full-frame DSLR to document a political rally, emphasizing crowd dynamics and key moments.

Applications of Digital Photography

Digital photography plays a pivotal role across numerous industries and creative fields. Its versatility, ease of use, and ability to produce high-quality visuals make it indispensable for both personal and professional applications. From preserving memories to shaping narratives, digital photography has redefined how we document, interpret, and share the world around us. Below are some of the most prominent applications of digital photography, each showcasing its ability to inform, inspire, and engage audiences.

1. Portrait Photography

Portrait photography focuses on capturing individuals or groups, emphasizing expressions, emotions, and personalities. This genre is widely used in both personal and professional contexts, offering timeless representations of people.

- **Use Cases**:
  - Professional headshots for resumes, portfolios, and LinkedIn profiles.
  - Family portraits for personal albums or holiday cards.
  - Creative portraits for art exhibitions or promotional campaigns.
- **Tools and Techniques**:
  - Cameras: DSLR and mirrorless cameras with wide-aperture lenses.
  - Lenses: Prime lenses like the 50mm or 85mm for sharp, detailed portraits with blurred backgrounds.
  - Lighting: Use of ring lights, softboxes, and reflectors to achieve flattering and even lighting.

- **Example**:
    - A corporate executive opts for a professional headshot taken in a studio with neutral backdrops, highlighting their personality and confidence.

2. Product Photography

Product photography is essential in the commercial and e-commerce sectors, where visual appeal significantly influences consumer behavior. High-quality images of products can make or break sales, particularly in online marketplaces.

- **Use Cases**:
    - E-commerce listings for platforms like Amazon, Etsy, and Shopify.
    - Advertisements and brochures for promoting new products.
    - Catalogs showcasing a company's product line.
- **Tools and Techniques**:
    - Cameras: High-resolution cameras that capture fine details.
    - Lenses: Macro lenses for close-ups and standard lenses for overall product shots.
    - Lighting: Lightboxes, diffusers, and studio setups to ensure even lighting without shadows.
    - Backgrounds: Clean, white backdrops for a professional look or contextual settings for lifestyle photography.
- **Example**:
    - A jewelry brand uses macro photography to highlight the intricate craftsmanship of a diamond necklace, ensuring every detail is visible and appealing to potential buyers.

3. Travel and Landscape Photography

Travel and landscape photography celebrates the beauty of natural and urban environments, offering visual narratives that inspire wanderlust and document diverse cultures and places. This genre often combines technical precision with artistic creativity to capture breathtaking scenes.

- **Use Cases**:
    - Editorial features in travel magazines and blogs.
    - Tourism campaigns promoting destinations.
    - Personal projects documenting adventures and experiences.
- **Tools and Techniques**:
    - Cameras: Lightweight mirrorless cameras for portability during travel.
    - Lenses: Wide-angle lenses for expansive landscapes and telephoto lenses for capturing distant subjects.
    - Techniques: Long-exposure photography to capture motion (e.g., waterfalls or clouds)

and HDR (High Dynamic Range) for balanced lighting in challenging conditions.

- **Example**:
  - A National Geographic photographer captures the vibrant hues of a desert sunset, using a wide-angle lens to include both the dramatic sky and the intricate patterns of sand dunes.

4. Event Photography

Event photography captures the essence and energy of various occasions, from personal milestones to professional gatherings. By documenting key moments, event photography ensures that memories are preserved and shared.

- **Use Cases**:
  - Weddings, engagements, and anniversaries for personal keepsakes.
  - Corporate events like conferences, product launches, and award ceremonies.
  - Concerts and performances to showcase the atmosphere and artistry.
- **Tools and Techniques**:
  - Cameras: DSLRs or mirrorless cameras with fast autofocus and high ISO performance for low-light conditions.
  - Lenses: Zoom lenses like the 24-70mm for versatility in dynamic settings.
  - Techniques: Use of burst mode to capture fleeting moments and careful composition to frame subjects against their environments.
- **Example**:
  - A wedding photographer uses a telephoto lens to discreetly capture an emotional moment between the bride and groom during the ceremony.

5. Photojournalism

Photojournalism uses powerful imagery to tell stories and document events, often serving as a visual complement to written reports. This genre emphasizes authenticity and immediacy, capturing moments that inform, move, and inspire audiences.

- **Use Cases**:
  - News coverage of political events, natural disasters, and social movements.
  - Editorial features in newspapers, magazines, and online platforms.
  - Documentary projects exploring cultural, historical, or humanitarian themes.
- **Tools and Techniques**:
  - Cameras: Rugged cameras capable of withstanding harsh conditions, such as the Canon EOS-1D X Mark III.
  - Lenses: Versatile zoom lenses for quick framing adjustments and prime lenses for low-light environments.

  - Techniques: Shooting in natural light to maintain authenticity and capturing candid moments to reflect reality.

- **Example**:
  - A Pulitzer Prize-winning photograph captures the resilience of a community rebuilding after a natural disaster, evoking both the tragedy and the hope of the moment.

The Broader Impact of Digital Photography

The applications of digital photography extend beyond individual categories, often overlapping and contributing to broader societal, artistic, and economic impacts. For instance:

- **Marketing and Advertising**:
  - High-quality images are critical for creating brand identities and engaging audiences.
  - Example: Luxury brands like Gucci and Rolex use professional photography to convey elegance and exclusivity.
- **Education and Research**:
  - Educational materials rely on clear and informative visuals to enhance learning experiences.
  - Example: Wildlife photographers document species for conservation projects and scientific studies.
- **Social Media and Content Creation**:
  - Platforms like Instagram, Pinterest, and Facebook thrive on visually appealing content, enabling individuals and brands to connect with audiences worldwide.
  - Example: Influencers use photography to curate aesthetic feeds and promote products.
- **Healthcare and Forensics**:
  - Medical imaging and forensic photography contribute to diagnostics, legal investigations, and evidence preservation.
  - Example: Dentists use macro photography to document patients' teeth before and after treatments.

The Synergy Between Digital Videography and Photography

Digital videography and photography are distinct yet complementary art forms, often intersecting to enhance storytelling and creative expression. Both mediums rely on similar technological foundations, such as cameras, lenses, and editing software, which allow creators to seamlessly integrate still images and moving visuals into cohesive narratives. This synergy amplifies the impact of multimedia content, providing a dynamic blend of motion, emotion, and detail. Modern digital cameras, with their ability to capture both high-resolution stills and cinematic video, have bridged the gap between videography and photography. This integration empowers creators to experiment with both mediums within the same project, maximizing their storytelling potential.

How Digital Videography and Photography Complement Each Other

The intersection of videography and photography enriches creative projects, offering unique advantages for various industries and applications. While photography freezes a single moment in time, videography captures motion, sound, and sequences. Together, they create a versatile toolkit for visual storytelling.

Key Areas of Synergy

1. Marketing Campaigns

In advertising and marketing, the combination of videography and photography allows brands to connect with diverse audiences across multiple platforms. The interplay between the two mediums ensures that campaigns are both visually appealing and strategically effective.

- **Photographic Use**:
  - High-quality photos are used in print materials such as brochures, billboards, and magazine ads.
  - Product images highlight details, texture, and features to attract consumers.
- **Videographic Use**:
  - Videos add motion, sound, and narrative to product demonstrations, brand stories, and customer testimonials.
  - Platforms like Instagram Reels, YouTube, and TikTok rely heavily on video content for audience engagement.
- **Synergy in Campaigns**:
  - Advertisers pair photographs for static promotions with videos for dynamic, interactive content.
  - Example: A luxury car brand uses still images to showcase design details in magazines and follows up with video ads that capture the car in motion, highlighting its performance and features.

2. Documentary Production

Documentary filmmakers frequently combine videography and photography to provide a more comprehensive narrative. While videos capture the flow of events and interviews, still photographs serve as poignant visual pauses, emphasizing critical moments or details.

- **Role of Photography**:
  - Still images document key moments that might otherwise be lost in the flow of a video.
  - These photographs are often used in promotional materials, posters, and accompanying books.
- **Role of Videography**:

    - Videos capture dynamic scenes, dialogues, and environmental sounds that bring the documentary to life.
    - Moving visuals provide context, motion, and emotional resonance.
- **Synergy in Documentaries**:
    - Photography and videography work together to build a layered narrative.
    - Example: A wildlife documentary might feature close-up photographs of an endangered species alongside video footage of its behavior in the wild, creating an emotional connection with viewers.

3. Social Media Content

Social media platforms thrive on visually driven content, making the integration of videography and photography a natural fit. Influencers, brands, and creators leverage both mediums to engage their audiences in diverse and impactful ways.

- **Photographic Use**:
    - Photos are curated for grid layouts, profile images, and thumbnails, offering quick and striking visuals.
    - Example: Instagram posts often feature high-resolution, aesthetic photos that reflect a personal or brand identity.
- **Videographic Use**:
    - Videos capture dynamic moments, tutorials, and storytelling sequences that draw viewers' attention.
    - Example: TikTok and Instagram Reels focus on short, engaging video formats to maximize reach and virality.
- **Synergy in Social Media**:
    - Photos establish the creator's visual identity, while videos provide a deeper, interactive experience.
    - Example: A travel influencer posts scenic photographs of a destination on Instagram while sharing immersive videos of the journey on YouTube, appealing to different audience preferences.

Technological Advancements Enabling Synergy

Advances in camera technology, editing software, and online platforms have made it easier than ever to blend photography and videography. Key innovations include:

- **Hybrid Cameras**:
    - Modern cameras like the Sony Alpha series and Canon EOS R series allow seamless transitions between stills and video.
    - Features such as high-resolution sensors, advanced autofocus, and dual shooting modes cater to both photographers and videographers.

- **Unified Editing Software**:
  - Tools like Adobe Lightroom, Photoshop, and Premiere Pro integrate photo and video editing, enabling creators to work across mediums within a single workflow.
- **AI Enhancements**:
  - Artificial intelligence streamlines editing processes, such as applying consistent color grading across photos and videos for a cohesive aesthetic.

Benefits of Integrating Videography and Photography

The synergy between videography and photography offers several advantages for creators and industries:

1. **Enhanced Storytelling**:
   - The combination of still and moving visuals enriches the narrative, adding layers of depth and emotion.
   - Example: A wedding photographer might include a slideshow of still photos alongside a video montage to create a comprehensive keepsake.
2. **Versatility**:
   - The ability to work in both mediums provides creators with greater flexibility to meet client demands and adapt to project requirements.
   - Example: A single camera operator can capture photos for promotional materials and videos for a corporate presentation.
3. **Audience Engagement**:
   - Using both mediums increases the reach and impact of content, catering to different audience preferences.
   - Example: Brands use videos to draw attention and photos to sustain interest through detailed exploration.
4. **Cost and Time Efficiency**:
   - Shooting both photos and videos during the same session minimizes production costs and streamlines workflows.
   - Example: A fashion shoot produces campaign photos and a behind-the-scenes video in one session, maximizing output.

Technological Advancements in Digital Videography and Photography

The fields of digital videography and photography have experienced transformative growth, fueled by cutting-edge technological advancements. These innovations have redefined creative possibilities, expanded accessibility, and elevated the quality of visual content. From artificial intelligence to aerial drones, the tools and techniques now available enable creators to produce professional-grade videos and photographs that captivate audiences across platforms.

1. AI and Machine Learning

Artificial intelligence (AI) and machine learning (ML) are revolutionizing videography and photography by automating complex tasks, enhancing precision, and streamlining workflows. These technologies are integrated into both hardware and software, empowering creators with advanced tools that simplify and accelerate editing processes.

- **Key Features**:
  - **Background Removal**: AI tools can identify and remove unwanted elements from images and videos, saving hours of manual editing.
    - Example: Adobe Photoshop's **Content-Aware Fill** allows users to remove distractions with a single click, seamlessly filling in the gaps.
  - **Auto-Tagging and Organization**: AI-powered platforms like Google Photos and Adobe Lightroom use facial recognition and object detection to categorize and organize visual content automatically.
  - **Dynamic Adjustments**: AI optimizes exposure, contrast, and color balance in real time, ensuring professional-quality results even for beginners.
- **Applications**:
  - Professional photographers use AI-enhanced editing to refine portraits, removing blemishes and enhancing details effortlessly.
  - Videographers rely on AI tools for motion tracking, automated color grading, and scene recognition, streamlining post-production.
- **Example**:
  - AI-driven tools like **Topaz Gigapixel AI** upscale low-resolution images and videos without sacrificing quality, making it ideal for restoring archival footage or repurposing older content.

2. 4K and 8K Technology

High-resolution technology has become the gold standard in videography and photography, offering unparalleled clarity, detail, and realism. **4K (3840x2160)** and **8K (7680x4320)** resolutions provide creators with the ability to capture intricate details, making these formats indispensable for professional projects.

- **Advantages**:
  - **Stunning Detail**: 4K and 8K resolutions allow for sharp, vibrant visuals that retain their quality even on large screens or during post-production cropping.
  - **Future-Proofing**: Content produced in 4K or 8K remains relevant as display technology continues to evolve.
  - **Flexibility in Editing**: High resolutions enable zooming, cropping, and stabilization without noticeable loss of quality.

- **Applications:**
    - Cinematic productions use 8K technology to achieve hyper-realistic visuals for theaters.
    - Advertising campaigns leverage 4K footage for immersive online and television promotions.
- **Example:**
    - The **RED Monstro 8K camera** is widely used in Hollywood, capturing visuals with extraordinary detail for blockbuster films.

3. Mobile Photography and Videography

Smartphone technology has democratized photography and videography, enabling millions of users to create high-quality content. Modern smartphones rival professional cameras in features and performance, making them versatile tools for creators on the go.

- **Key Features:**
    - **Multiple Lenses**: Smartphones like the Apple iPhone Pro series and Samsung Galaxy Ultra feature wide-angle, telephoto, and macro lenses, offering flexibility in capturing diverse scenes.
    - **AI Enhancements**: Integrated AI improves image stabilization, low-light performance, and real-time scene recognition.
    - **RAW Image Capture**: Advanced smartphones support RAW formats, allowing for greater control over editing and post-processing.
    - **HDR Video Recording**: High Dynamic Range (HDR) enhances contrast and color, producing vibrant and lifelike visuals.
- **Applications:**
    - Social media influencers and content creators use smartphones for producing short-form videos and stunning photography for platforms like Instagram and TikTok.
    - News organizations rely on mobile videography for real-time reporting, capturing events as they unfold.
- **Example:**
    - Apple's **iPhone Pro series** supports **Dolby Vision HDR** video recording, enabling cinematic-quality videos directly from a smartphone.

4. Drones

Drones have revolutionized aerial photography and videography, offering unique perspectives that were once only possible with helicopters or cranes. Equipped with advanced stabilization and high-resolution cameras, drones are now essential tools in industries ranging from filmmaking to real estate.

- **Key Features**:
    - **GPS and Navigation Systems**: Enable precise positioning and automated flight paths for complex shots.
    - **Obstacle Avoidance**: Ensures safe operation in challenging environments.
    - **High-Resolution Cameras**: Many drones now support 4K and even 8K video recording, as well as RAW photography.
- **Applications**:
    - **Filmmaking**: Drones provide sweeping aerial shots, adding cinematic flair to movies and documentaries.
    - **Real Estate Marketing**: Aerial visuals showcase properties from unique angles, highlighting their location and surroundings.
    - **Event Coverage**: Weddings and sports events use drones for dynamic, bird's-eye-view footage.
- **Example**:
    - The **DJI Phantom 4 Pro** is a popular choice for professional drone videography, featuring a 20MP camera and 4K recording at 60 fps.

5. 360-Degree Cameras

360-degree cameras capture a full spherical view of a scene, enabling immersive content creation for virtual reality (VR), augmented reality (AR), and interactive storytelling. These cameras are gaining traction in industries that prioritize engagement and innovation.

- **Key Features**:
    - **Omnidirectional Lenses**: Capture every angle simultaneously, creating a seamless panoramic view.
    - **Stitching Technology**: Combines footage from multiple lenses into a single, cohesive image or video.
    - **Compact Design**: Portable and lightweight, making 360-degree cameras ideal for on-the-go creators.
- **Applications**:
    - **Virtual Tours**: Real estate agencies and museums use 360-degree cameras to offer immersive walkthroughs of properties and exhibitions.
    - **VR Content**: Gaming and entertainment industries leverage 360-degree footage to create lifelike virtual experiences.
    - **Adventure Filmmaking**: Action sports enthusiasts use 360-degree cameras to capture dynamic footage from unique perspectives.

- **Example**:
    - The **Insta360 ONE X2** is a leading 360-degree camera, offering 5.7K resolution and advanced stabilization for professional-quality immersive content.

The Impact of Technological Advancements

These technological innovations have not only enhanced the quality and accessibility of videography and photography but have also transformed how visual content is created and consumed:

1. **Increased Accessibility**:
    - Affordable tools like smartphones and drones have lowered the entry barriers, enabling creators from diverse backgrounds to participate in the visual arts.
2. **Creative Freedom**:
    - AI-driven editing and high-resolution capabilities give creators greater flexibility and precision in realizing their artistic visions.
3. **Engaging Content**:
    - Immersive technologies like 360-degree cameras and VR enable audiences to experience content in new and interactive ways.
4. **Industry Growth**:
    - Advancements in technology have fueled the growth of industries like e-commerce, entertainment, and real estate, where high-quality visuals are essential for success.

C. Digital Text Writing

Digital text writing is a cornerstone of modern communication, encompassing the creation, editing, and dissemination of written content through digital platforms. This form of writing transcends traditional pen-and-paper methods, leveraging technology to produce content that is accessible, interactive, and versatile. From crafting blogs and social media posts to producing e-books and academic articles, digital textwriting has become integral to personal expression, professional work, and global connectivity. Unlike traditional writing, digital text writing offers unique features such as hyperlinks, multimedia integration, and instant distribution. These attributes not only enhance the reader's experience but also empower writers to reach diverse audiences in innovative ways. The dynamic nature of digital text writing makes it adaptable across industries, including journalism, marketing, education, and creative arts.

1. Key Components of Digital Text Writing

Digital text writing relies on a combination of advanced tools, refined techniques, and strategic approaches to produce content that is engaging, accessible, and impactful. These components work together to streamline the writing process, improve quality, and enable writers to craft content that resonates with diverse audiences. The ability to leverage these tools and strategies effectively is essential for success in the fast-paced digital world.

1. Tools for Digital Text Writing

Modern technology has revolutionized how writers create, edit, and distribute content. These tools not only simplify the writing process but also enable collaboration, improve accuracy, and enhance creativity. From general-purpose applications to specialized software, digital tools have become indispensable for writers across various domains.

Word Processors

Word processors are the most basic yet essential tools for digital text writing. They provide a versatile platform for creating, editing, and formatting text.

- **Popular Tools**:
  - **Microsoft Word**: A longstanding leader in text editing, known for its robust formatting options and compatibility.
  - **Google Docs**: A cloud-based platform that facilitates real-time collaboration and access across devices.
  - **Apple Pages**: Tailored for macOS users, combining simplicity with powerful design features.
- **Benefits**:
  - **Real-Time Collaboration**: Platforms like Google Docs allow multiple users to work on a document simultaneously.
  - **Auto-Saving**: Automatic saving ensures that work is never lost due to unexpected interruptions.
  - **Accessibility**: Cloud integration enables writers to access their documents from any device with an internet connection.
- **Example**:
  - A content marketing team uses Google Docs to draft, review, and finalize blog posts collaboratively, streamlining their workflow.

Content Management Systems (CMS)

Content Management Systems (CMS) are platforms that enable writers to publish, organize, and manage online content efficiently. They are particularly valuable for creating websites, blogs, and e-commerce pages.

- **Popular Platforms**:
  - **WordPress**: A highly customizable CMS used by bloggers, businesses, and organizations worldwide.
  - **Squarespace**: Known for its sleek templates and ease of use, ideal for small businesses and creatives.
  - **Wix**: Offers drag-and-drop functionality, making it accessible to users with minimal technical expertise.

- **Benefits**:
    - **User-Friendly Interfaces**: Simplify the process of publishing and organizing content.
    - **Built-In SEO Tools**: Help optimize content for search engines, increasing visibility and reach.
    - **Customization**: Allow users to design their websites and content layouts according to their branding needs.
- **Example**:
    - A freelance writer uses WordPress to host their portfolio, publish blogs, and attract potential clients.

Writing Enhancement Tools

Writing enhancement tools are designed to refine the quality of content by addressing grammar, style, and readability. These tools are essential for producing polished and professional work.

- **Popular Applications**:
    - **Grammarly**: Checks grammar, spelling, and punctuation while offering tone and style suggestions.
    - **ProWritingAid**: Focuses on improving overall readability, including sentence structure and clarity.
    - **Hemingway Editor**: Highlights overly complex sentences and passive voice, promoting simplicity and engagement.
- **Benefits**:
    - **Instant Feedback**: Provides real-time suggestions to correct errors and improve quality.
    - **Tone Analysis**: Ensures the writing aligns with the intended audience and purpose.
    - **Clarity Improvements**: Helps writers identify and resolve areas of ambiguity.
- **Example**:
    - An academic writer uses Grammarly to ensure their research paper meets professional standards of grammar and style.

Specialized Software

Specialized software caters to specific types of writing, such as novel creation, screenplay formatting, and technical documentation. These tools provide features tailored to the unique requirements of particular writing projects.

- **Popular Tools**:
    - **Scrivener**: Designed for authors, Scrivener offers tools for organizing chapters, research notes, and drafts.

    - **Final Draft**: The industry standard for screenwriting, offering precise formatting and collaboration tools.
    - **MadCap Flare**: Used for creating technical documentation and user manuals with advanced formatting options.
- **Benefits**:
    - **Streamlined Workflows**: Provide templates and organization features that simplify complex projects.
    - **Collaboration**: Enable teams to work together efficiently on large-scale writing endeavors.
- **Example**:
    - A novelist uses Scrivener to organize their manuscript, research, and character development notes in a single workspace.

Speech-to-Text Technology

Speech-to-text technology allows writers to dictate their content, converting spoken words into text. This alternative to traditional typing is particularly useful for improving productivity and accommodating different working styles.

- **Popular Programs**:
    - **Dragon NaturallySpeaking**: Known for its accuracy and customization options.
    - **Google Voice Typing**: A free tool integrated into Google Docs for basic dictation needs.
    - **Otter.ai**: Combines transcription with AI-driven insights, making it ideal for meetings and interviews.
- **Benefits**:
    - **Increased Speed**: Enables writers to produce content faster by dictating rather than typing.
    - **Accessibility**: Assists individuals with physical disabilities or repetitive strain injuries.
    - **Flexibility**: Allows for idea capture during brainstorming sessions or while on the move.
- **Example**:
    - A journalist uses Otter.ai to transcribe interviews, saving time and ensuring accuracy in reporting.

2. Key Components of Digital Text Writing

Digital textwriting is a versatile craft shaped by a blend of tools, techniques, and styles. Understanding its key components is essential for producing content that resonates with audiences and achieves its intended purpose. These components encompass the technologies that facilitate writing, the stylistic approaches tailored to specific audiences, and the strategic considerations that make digital text impactful.

1. Tools for Digital Text Writing

Modern technology offers writers a suite of tools to simplify and enhance the process of creating digital content. These tools are designed to improve efficiency, ensure quality, and foster creativity.

1.1 Word Processors

Word processors are foundational tools for text creation, offering features that streamline drafting, editing, and formatting.

- **Popular Options**:
    - **Microsoft Word**: Known for its comprehensive features and compatibility across devices.
    - **Google Docs**: Offers cloud-based collaboration with real-time editing.
    - **Apple Pages**: Combines user-friendly design with advanced writing features for Mac users.
- **Benefits**:
    - Real-time collaboration: Enables multiple users to work on the same document simultaneously.
    - Auto-saving: Minimizes the risk of losing work.
    - Accessibility: Allows writers to access documents across various devices.
- **Example**:
    - A freelance writer drafts a blog post in Google Docs, collaborating with an editor to finalize content for publication.

1.2 Content Management Systems (CMS)

Content Management Systems are platforms that help writers publish, organize, and manage digital content on websites.

- **Popular CMS Platforms**:
    - **WordPress**: Powers a significant percentage of websites, offering extensive customization and plugins.
    - **Squarespace**: Known for its sleek, professional templates.
    - **Wix**: Provides a user-friendly drag-and-drop interface.
- **Benefits**:
    - Built-in SEO tools: Optimize content for search engines.
    - Design customization: Allows writers to align content presentation with brand identity.

- Analytics integration: Tracks audience engagement and performance metrics.

- **Example**:
    - A small business owner uses WordPress to create a blog that complements their online store, integrating SEO and analytics tools to attract traffic.

1.3 Writing Enhancement Tools

These tools refine the quality of written content by providing feedback on grammar, style, tone, and readability.

- **Popular Applications**:
    - **Grammarly**: Highlights grammatical errors, tone inconsistencies, and style issues.
    - **ProWritingAid**: Offers in-depth analysis for fiction and long-form writing.
    - **Hemingway Editor**: Simplifies complex sentences to improve readability.
- **Benefits**:
    - Instant feedback: Reduces editing time by identifying issues during the writing process.
    - Tone analysis: Ensures content aligns with the intended audience.
    - Style suggestions: Improves clarity and flow.
- **Example**:
    - A marketer uses Grammarly to polish an email campaign, ensuring professional and engaging communication.

1.4 Specialized Software

Tailored for specific writing needs, specialized software enhances productivity and organization in niche fields.

- **Popular Software**:
    - **Scrivener**: Designed for authors, Scrivener offers tools for structuring novels, screenplays, and research projects.
    - **Final Draft**: A standard in the film industry, used for creating and formatting screenplays.
- **Example**:
    - A novelist organizes plotlines and character development using Scrivener's visual storyboard features.

1.5 Speech-to-Text Technology

Speech-to-text tools convert spoken words into written text, providing an alternative to traditional typing.

- **Popular Programs**:
  - **Dragon NaturallySpeaking**: Known for its accuracy and advanced customization.
  - **Google Voice Typing**: A free option integrated with Google Docs.
- **Benefits**:
  - Accessibility: Supports writers with physical disabilities or those who prefer dictation.
  - Speed: Accelerates the drafting process, particularly for brainstorming or long-form content.
- **Example**:
  - A journalist uses Dragon NaturallySpeaking to transcribe interview recordings, saving time during the research phase.

2. Writing Styles in Digital Text

Digital platforms have transformed the way content is written and consumed, leading to the emergence of diverse writing styles. Each style is tailored to specific audiences and objectives, ensuring that the message resonates effectively.

2.1 Conversational Writing

Conversational writing mimics natural speech, creating a friendly and relatable tone. It is particularly effective for engaging audiences in informal contexts.

- **Characteristics**:
  - Use of first-person and second-person perspectives.
  - Simple, concise language with contractions and colloquialisms.
- **Applications**:
  - Blogs, social media posts, and lifestyle articles.
- **Example**:
  - A travel blogger shares personal experiences with an approachable tone: "Imagine sipping coffee while watching the sunrise over the mountains—it's pure magic."

2.2 Professional Writing

Professional writing emphasizes clarity, structure, and objectivity, making it suitable for formal contexts.

- **Characteristics**:
  - Adherence to industry-specific jargon and terminology.
  - Structured formats, such as bullet points, headings, and tables.
- **Applications**:
  - Business reports, academic papers, and corporate communications.

- **Example:**
    - A whitepaper outlines the benefits of cloud computing with precise, data-backed arguments: "Cloud-based solutions improve operational efficiency by 30% while reducing costs."

2.3 Persuasive Writing

Persuasive writing aims to influence opinions or encourage specific actions, making it central to marketing and advertising.

- **Characteristics:**
    - Use of compelling language and emotional appeals.
    - Clear call-to-action (CTA) statements.
- **Applications:**
    - E-commerce product descriptions, promotional emails, and advocacy campaigns.
- **Example:**
    - A product page description highlights benefits: "Unlock your creativity with our state-of-the-art graphic tablet—perfect for artists, designers, and innovators."

2.4 Creative Writing

Creative writing showcases originality and emotional depth, often prioritizing artistic expression over functionality.

- **Characteristics:**
    - Rich imagery, metaphors, and storytelling elements.
    - Flexible structure and style.
- **Applications:**
    - Fiction, poetry, and narrative nonfiction.
- **Example:**
    - A short story on a digital platform like Wattpad captures the reader's imagination: "The stars whispered secrets only she could hear as she wandered the forest paths."

3. Formatting and Design in Digital Text Writing

Formatting and design are critical aspects of digital text writing that go beyond the words themselves. These elements play a significant role in enhancing readability, user engagement, and the overall aesthetic appeal of content. Thoughtful formatting ensures that readers can navigate and comprehend information effortlessly, while design elements make the content visually engaging and interactive. Together, they transform written content into an immersive digital experience. Effective formatting and design are particularly important in the digital age, where audiences consume content on a variety of devices with differing screen sizes and resolutions. Whether crafting a blog post, an e-book, or a web page, writers must balance visual appeal with functionality, ensuring that the content meets the needs of diverse audiences.

Key Elements of Formatting and Design

1. Typography

Typography is the foundation of digital text design, encompassing the choice of fonts, font sizes, spacing, and alignment. It directly impacts how readers perceive and engage with the content.

- **Font Choice**:
  - Readable and visually appealing fonts enhance user experience.
  - **Sans-serif fonts**, such as Arial, Helvetica, and Roboto, are preferred for digital content due to their clean, modern look and high legibility on screens.
  - **Serif fonts**, like Times New Roman and Georgia, are more traditional and often used for printed text or formal digital content.
- **Font Size**:
  - Optimal font size ensures readability across devices.
  - Example: A standard web font size of **16px** is widely considered comfortable for on-screen reading.
- **Line Spacing and Alignment**:
  - Adequate spacing between lines and paragraphs prevents text from feeling cramped.
  - Left-aligned text is the norm for digital content, ensuring a consistent reading flow.
- **Example in Practice**:
  - A blog about health tips uses Roboto in 16px for body text, with headings in a larger, bold sans-serif font like Open Sans for clear section differentiation.

2. Hyperlinks

Hyperlinks are an essential feature of digital text writing, enabling interactivity and providing additional value to readers by connecting them to related resources.

- **Types of Hyperlinks**:
  - **Internal Links**: Direct readers to other pages or posts within the same website, improving site navigation and encouraging engagement.
    - Example: A fitness blog links to its recipe section from an article on healthy eating.
  - **External Links**: Point to authoritative sources or references outside the website, enhancing credibility and context.

- **Best Practices for Hyperlinking**:
    - Use descriptive anchor text that clearly indicates where the link will lead.
        - Example: Instead of "click here," use "Learn more about digital marketing strategies."
    - Avoid overuse of links, as excessive hyperlinking can distract readers.
- **Impact on SEO**:
    - Proper use of hyperlinks boosts search engine rankings by improving site structure and providing value to readers.

3. Multimedia Integration

Incorporating multimedia elements, such as images, videos, and infographics, enriches digital content by breaking up text-heavy sections and offering visual context.

- **Images**:
    - Relevant and high-quality images enhance understanding and retention of information.
    - Example: A blog on travel destinations includes scenic photographs of the locations described.
- **Videos**:
    - Videos engage audiences by presenting complex concepts in an accessible and entertaining format.
    - Example: An instructional blog embeds a video tutorial alongside step-by-step written instructions.
- **Infographics**:
    - Simplify and summarize data using visually appealing graphics.
    - Example: A financial blog includes an infographic illustrating budgeting tips.
- **Alt Text**:
    - Descriptive alternative text for images improves accessibility for visually impaired users and contributes to SEO.

4. Interactive Features

Interactive elements transform static text into an engaging experience, encouraging readers to participate actively with the content.

- **Embedded Forms**:
    - Collect reader input for surveys, feedback, or subscriptions.
    - Example: A newsletter signup form embedded at the end of a blog post.

- **Quizzes and Polls**:
  - Encourage audience engagement by prompting readers to interact with the content.
  - Example: A personality quiz within a lifestyle blog.
- **Comment Sections**:
  - Foster community interaction and discussion, allowing readers to share thoughts and ask questions.
  - Example: A tech blog includes a comment section where readers can discuss the latest product reviews.
- **Interactive Infographics**:
  - Allow users to explore data by clicking or hovering over elements.
  - Example: An educational website provides an interactive map of historical events.

Why Formatting and Design Matter

1. **Improved Readability**:
   - Proper formatting ensures that content is easy to read, even for audiences scanning quickly on mobile devices.
2. **Enhanced Engagement**:
   - Multimedia and interactive features capture and retain reader attention, encouraging longer time spent on the page.
3. **Accessibility**:
   - Thoughtful design makes content accessible to a wider audience, including individuals with visual or cognitive impairments.
4. **Professionalism**:
   - A well-designed piece of digital text reflects credibility and attention to detail, reinforcing the author's or organization's authority.
5. **SEO Benefits**:
   - Search engines prioritize well-structured, visually appealing content that provides a positive user experience.

Best Practices for Digital Formatting and Design

- Use headings (H1, H2, H3) to structure content logically, making it easy to navigate.
- Break up large paragraphs into smaller chunks for readability.
- Use bullet points or numbered lists to present information clearly and concisely.
- Include a mix of visual and textual elements to cater to different learning preferences.
- Optimize multimedia files to ensure fast loading times and improve user experience.

4. Search Engine Optimization (SEO)

Search Engine Optimization (SEO) is a fundamental element of digital text writing, enabling content to rank higher in search engine results pages (SERPs) and reach a broader audience. In a digital landscape dominated by search engines like Google, Bing, and Yahoo, mastering SEO is essential for ensuring that content remains visible, competitive, and impactful. Effective SEO strategies combine technical precision, creative writing, and user-centric design to enhance the discoverability and relevance of digital content.

Key Components of SEO

1. Keywords

Keywords are the cornerstone of SEO, representing the search terms that users enter into search engines. Strategic placement of relevant keywords throughout the text helps search engines understand the content's purpose and match it to user queries.

- **Best Practices for Keyword Use**:
    - Conduct keyword research using tools like Google Keyword Planner, Ahrefs, or SEMrush to identify terms with high search volume and low competition.
    - Incorporate primary keywords naturally into the title, headings, and body of the text.
    - Use long-tail keywords—specific phrases that target niche audiences for improved precision.
- **Keyword Density**:
    - Avoid keyword stuffing, which can lead to penalties from search engines. A keyword density of 1–2% is generally acceptable.
- **Example**:
    - A blog on digital marketing might feature keywords like "SEO strategies," "content marketing tips," and "search engine ranking."

2. Meta Descriptions

Meta descriptions are short summaries that appear below the page title in search results. While not a direct ranking factor, they significantly impact click-through rates (CTR) by enticing users to visit the page.

- **Characteristics of Effective Meta Descriptions**:
    - Concise: Ideally between 150–160 characters.
    - Informative: Clearly convey the page's content and value.
    - Action-Oriented: Include a call-to-action (CTA) to encourage clicks.
- **Example**:
    - For an article titled "10 Tips for Improving Your SEO," the meta description

might read: "Discover proven strategies to boost your website's search rankings, increase traffic, and grow your online presence."

3. Headings and Subheadings

Headings and subheadings (H1, H2, H3, etc.) structure content logically, improving readability and making it easier for search engines to crawl and index the page.

- **Best Practices**:
    - Use an **H1 tag** for the main title and **H2 or H3 tags** for subsections.
    - Include keywords in headings to signal relevance to search engines.
    - Ensure that headings are descriptive and concise to guide readers effectively.
- **Example**:
    - In an article about social media marketing, headings might include:
        - H1: "Mastering Social Media Marketing"
        - H2: "Key Platforms for Social Media Success"
        - H3: "Tips for Engaging Content Creation"

4. Alt Text for Images

Alt text (alternative text) describes the content and purpose of images within a web page. It enhances accessibility for visually impaired users and improves SEO by providing search engines with additional context about the page.

- **Best Practices for Alt Text**:
    - Be descriptive and specific about the image.
    - Include keywords where appropriate, but avoid keyword stuffing.
    - Ensure alt text aligns with the page's overall content and message.
- **Example**:
    - An e-commerce site featuring a product image of a smartwatch might use alt text like: "Black smartwatch with fitness tracking features, compatible with iOS and Android."

Advanced SEO Techniques

In addition to these foundational components, advanced SEO techniques can further optimize digital text writing for search engines:

1. Internal and External Linking

- **Internal Links**:
    - Connect different pages within the same website to improve navigation and distribute link equity.

  - Example: A blog post on SEO links to a related article on keyword research.
- **External Links**:
  - Direct users to authoritative sources, enhancing credibility and providing additional value.
  - Example: Citing a reputable study on content marketing trends.

2. Mobile Optimization

Search engines prioritize mobile-friendly websites due to the growing number of users accessing content on smartphones and tablets.

- **Best Practices**:
  - Use responsive design to ensure content adapts to different screen sizes.
  - Optimize images and multimedia files for faster loading times.

3. Page Speed

Fast-loading pages improve user experience and SEO rankings.

- **Strategies**:
  - Compress images and videos to reduce file sizes.
  - Minimize the use of heavy scripts and plugins.

4. Schema Markup

Schema markup adds structured data to web pages, helping search engines understand the content better and display rich snippets in search results.

- **Example**:
  - A recipe blog uses schema markup to display cooking time, ingredients, and user ratings directly in the search results.

Benefits of SEO in Digital Text Writing

1. **Increased Visibility**:
   - SEO ensures content ranks higher on SERPs, making it more likely to attract organic traffic.
2. **Enhanced User Experience**:
   - Proper formatting, fast loading times, and mobile optimization improve accessibility and engagement.
3. **Higher Credibility**:
   - Appearing at the top of search results signals authority and reliability to users.
4. **Cost-Effectiveness**:
   - SEO generates organic traffic, reducing reliance on paid advertising.

5. **Long-Term Results:**
    - Unlike paid campaigns, SEO efforts yield lasting benefits as content continues to attract traffic over time.

Common SEO Mistakes to Avoid

- Keyword stuffing, which harms readability and can result in penalties.
- Ignoring mobile optimization, which alienates a significant portion of users.
- Neglecting meta descriptions, leading to missed opportunities for engagement.
- Failing to update outdated content, which can affect rankings negatively.

5. Applications of Digital Text Writing

Digital text writing is a cornerstone of modern communication, transcending traditional writing by leveraging technology to create content that is interactive, accessible, and far-reaching. Its versatility enables applications across diverse domains, including content creation, journalism, marketing, education, and the creative arts. By adapting to various contexts and utilizing digital tools, writers can craft impactful messages that resonate with their intended audiences.

1. Content Creation

Content creation represents one of the most dynamic applications of digital text writing, allowing individuals and organizations to communicate effectively and creatively with diverse audiences.

- **Blogs:**
    - Blogs serve as digital platforms for sharing insights, expertise, and stories on virtually any topic.
    - **Applications:**
        - Personal blogs: Writers share experiences, hobbies, and opinions.
        - Corporate blogs: Businesses use blogs to demonstrate thought leadership, explain products, and engage audiences.
    - **Example:**
        - A technology blog discusses advancements in artificial intelligence, linking to relevant research and multimedia demonstrations.
- **Social Media Posts:**
    - Concise, engaging posts are critical for building brand awareness and fostering interaction on platforms like Twitter, Instagram, and LinkedIn.
    - **Applications:**
        - Twitter threads summarize complex topics into digestible parts.
        - LinkedIn posts highlight professional achievements or industry insights.

- **Example**:
  - A tweet encapsulates key findings from a new marketing report, encouraging retweets and discussions.

- **E-Newsletters**:
  - Regular email newsletters keep subscribers informed about updates, promotions, and curated content.
  - **Applications**:
    - Businesses use newsletters to announce product launches or share industry insights.
    - Nonprofits send newsletters to update donors on their initiatives.
  - **Example**:
    - A fitness brand shares weekly workout tips, healthy recipes, and product promotions through a subscription-based newsletter.

2. Journalism

Digital text writing has transformed journalism, enabling real-time reporting, interactive storytelling, and broader audience reach. It caters to the fast-paced nature of news dissemination while allowing for in-depth exploration of issues.

- **Online News Articles**:
  - News websites use digital text to deliver breaking stories enriched with multimedia elements and hyperlinks for context.
  - **Applications**:
    - Embedding videos and photos for visual storytelling.
    - Linking to primary sources for greater credibility.
  - **Example**:
    - A New York Times article reports on climate change, incorporating interactive graphs and related articles.
- **Opinion Pieces**:
  - Digital platforms provide a space for thought leaders to share perspectives and generate discourse.
  - **Applications**:
    - Editorial columns on current events or policy changes.
    - Personal essays reflecting on societal trends.
  - **Example**:
    - A journalist writes an opinion piece on the ethical implications of artificial intelligence, sparking online debates.

- **Data Journalism**:
  - Combines text with data visualizations to present complex information in an accessible format.
  - **Applications**:
    - Interactive charts explain economic trends.
    - Maps illustrate election results by region.
  - **Example**:
    - A publication visualizes the impact of the COVID-19 pandemic using heatmaps and detailed statistics.

3. Marketing and Advertising

Digital text writing is central to marketing and advertising, crafting messages that resonate with target audiences and encourage action.

- **Website Copy**:
  - Clear and persuasive website content guides users through the customer journey, from discovery to conversion.
  - **Applications**:
    - Homepages highlight brand values and offerings.
    - Product pages emphasize features and benefits.
  - **Example**:
    - A software company's homepage outlines its platform's ease of use, cost-effectiveness, and customer testimonials.
- **Product Descriptions**:
  - Engaging descriptions highlight the value of products, influencing purchasing decisions.
  - **Applications**:
    - E-commerce platforms use detailed descriptions to inform and persuade buyers.
    - Luxury brands emphasize exclusivity and craftsmanship.
  - **Example**:
    - An online store describes a smartwatch's fitness tracking, water resistance, and compatibility with mobile devices.
- **Ad Campaigns**:
  - Short, compelling text in digital ads captures attention and conveys key messages instantly.

- **Applications**:
    - Banner ads entice clicks with catchy headlines.
    - Social media ads combine text with visuals to boost engagement.
- **Example**:
    - A Facebook ad for an eco-friendly product uses persuasive language like "Sustainable, Stylish, and Affordable—Shop Now!"

4. Education and Training

Digital text writing supports diverse educational needs by creating interactive, informative, and accessible learning materials.

- **E-Learning Modules**:
    - Interactive text, paired with visuals and quizzes, engages learners and enhances comprehension.
    - **Applications**:
        - Online courses on platforms like Coursera or Udemy.
        - Corporate training modules for employee development.
    - **Example**:
        - An e-learning module teaches coding basics with step-by-step instructions and interactive exercises.
- **Instruction Manuals**:
    - Digital guides offer clear, user-friendly instructions for using products or services.
    - **Applications**:
        - Software user manuals detail installation and troubleshooting.
        - Appliance guides explain operation and maintenance.
    - **Example**:
        - A PDF manual for a camera provides diagrams, FAQs, and troubleshooting tips.
- **Academic Writing**:
    - Digital formats expand access to research papers, essays, and textbooks.
    - **Applications**:
        - Universities publish open-access journals.
        - Students submit essays and dissertations online.
    - **Example**:
        - A research paper on climate policy is published in a digital journal, complete with citations and hyperlinks.

5. Creative Arts

Digital text writing enables creative expression, offering platforms for writers to publish and share their work globally.

- **Fiction and Poetry**:
    - Digital platforms host novels, short stories, and poems, reaching readers without the constraints of traditional publishing.
    - **Applications**:
        - Self-published e-books on Amazon Kindle.
        - Poetry shared on platforms like Medium.
    - **Example**:
        - An independent author releases a sci-fi novel as an e-book, gaining a dedicated readership online.
- **Interactive Storytelling**:
    - Digital tools blend text with multimedia elements, creating immersive narratives.
    - **Applications**:
        - Choose-your-own-adventure stories integrate clickable choices.
        - Augmented reality (AR) storytelling combines text with interactive visuals.
    - **Example**:
        - A historical fiction website allows readers to explore the story through timelines, maps, and character journals.

Advantages of Digital Text Writing

1. **Accessibility**:
    - Digital content is accessible on multiple devices, enabling readers to consume information anytime, anywhere.
2. **Interactivity**:
    - Features like hyperlinks, multimedia, and comment sections foster active reader participation.
3. **Scalability**:
    - Digital text can reach global audiences without the limitations of physical distribution.
4. **Cost-Effectiveness**:
    - Digital platforms eliminate printing and distribution expenses, reducing overall production costs.
5. **Real-Time Updates**:
    - Writers can instantly edit and update digital content, ensuring accuracy and relevance.

Chapter 7

HTML Document Structure

Introduction

The evolution of the internet has transformed how we create, share, and consume information, with web and blog designing emerging as fundamental tools in this digital age. From personal blogs and small business websites to complex e-commerce platforms and online communities, web design is the foundation of how individuals and organizations present themselves online. Effective web design combines technical skills, creativity, and user-focused strategies to create visually appealing, functional, and accessible online spaces.

At its core, web and blog designing is a multidisciplinary endeavor. It encompasses elements of coding, graphic design, user experience (UX), and search engine optimization (SEO), each playing a vital role in crafting an engaging digital presence. For beginners, understanding basic web technologies, such as HTML (HyperText Markup Language) and CSS (Cascading Style Sheets), is the first step toward building and customizing websites. These tools enable creators to design the structure, style, and layout of web pages while ensuring they are responsive, accessible, and user-friendly. Blogs, as a subset of web design, have gained immense popularity as platforms for personal expression, professional branding, and content marketing. Blogging allows writers to connect with audiences by sharing insights, stories, and expertise in a structured and interactive format. Modern blog platforms and content management systems (CMS) simplify the process, empowering even those with limited technical knowledge to create and manage their online presence effectively.

Hosting, another critical component of web and blog design, ensures that websites and blogs are accessible on the internet. Hosting services provide the infrastructure necessary to store website files, manage traffic, and deliver content to users. Choosing the right hosting service—whether shared, dedicated, or cloud-based—is a strategic decision that affects website performance, scalability, and security.

The role of HTML in web design cannot be overstated. HTML forms the backbone of every web page, defining its structure and content. Understanding basic tags and the HTML document structure enables creators to add elements such as text, images, links, and multimedia to their sites. Alongside HTML, CSS serves as the styling language, dictating the visual presentation of web pages, including colors, fonts, spacing, and layout. Together, HTML and CSS allow for seamless customization and optimization of web pages to meet aesthetic and functional goals. Cascading Style Sheets (CSS) elevate web design by separating content from presentation. This division not only simplifies design updates but also enhances consistency across a website. By leveraging CSS, designers can ensure that

DOI: 10.4324/9781003750444-7

their sites are responsive—adapting seamlessly to different devices and screen sizes. Advanced CSS techniques, such as animations and transitions, add an interactive layer to web pages, captivating users and enhancing their browsing experience.

Web and blog designing is more than just technical know-how; it is a medium of communication and creativity. It reflects the identity of the creator or organization while addressing the needs and expectations of its audience. As the digital landscape continues to evolve, mastering web and blog design skills remains essential for anyone seeking to establish a robust online presence, whether for personal branding, business growth, or creative expression.

A. Hosting; Introduction of HTML and Basic Tags and HTML Document Structure

Building a functional and engaging website or blog requires a clear understanding of two fundamental components: hosting and HTML. Hosting serves as the backbone of web accessibility, ensuring that websites and blogs are available to users around the world. HTML, on the other hand, is the foundation of website structure, dictating how content is displayed and interacted with. Together, they form the starting point for any web design project.

1. Hosting

Hosting is the backbone of any online presence, serving as the infrastructure that makes websites and blogs accessible to users around the globe. Without hosting, websites would exist only on local machines, invisible to the public. Hosting services ensure that website files are stored securely on servers, which are specialized computers designed to manage internet traffic, deliver content efficiently, and maintain high uptime rates. Choosing the right hosting service is critical, as it affects the performance, security, scalability, and user experience of a website. Different types of hosting cater to various needs, ranging from personal blogs to large-scale enterprise applications.

Types of Hosting

Web hosting is not a one-size-fits-all solution. The choice of hosting depends on factors such as traffic levels, the complexity of the website, and the budget. Below are the primary types of hosting, each with unique features, advantages, and limitations.

1. Shared Hosting

Shared hosting is one of the most popular and affordable hosting solutions, making it an excellent choice for beginners and small-scale projects.

- **How It Works**:
  - Multiple websites share the resources of a single server, including CPU, memory, and storage.
  - Hosting providers allocate specific amounts of server resources to each website.
- **Ideal For**:
  - Personal blogs, portfolios, and small business websites with low to moderate traffic.

- **Pros**:
  - **Cost-Effective**: Shared hosting plans are typically the least expensive, with some starting as low as a few dollars per month.
  - **User-Friendly**: Hosting providers offer intuitive control panels like cPanel, making it easy for non-technical users to manage their websites.
  - **Pre-Configured**: Servers are pre-configured, reducing setup time.
- **Cons**:
  - **Performance Limitations**: High traffic on one website can impact the performance of others on the same server.
  - **Limited Control**: Users have restricted access to server configurations.
  - **Security Risks**: Shared environments may pose risks if one website is compromised.
- **Example Providers**:
  - Bluehost, HostGator, and SiteGround.

2. Dedicated Hosting

Dedicated hosting provides exclusive access to an entire server for a single website, delivering unmatched performance and control.

- **How It Works**:
  - A single server is leased by one client, who has complete control over its configuration and resources.
- **Ideal For**:
  - Large businesses, e-commerce platforms, and websites with high traffic volumes requiring robust performance and security.
- **Pros**:
  - **High Performance**: Resources are not shared, ensuring consistent speed and reliability.
  - **Enhanced Security**: Dedicated environments minimize risks of breaches caused by neighboring websites.
  - **Full Customization**: Users can configure the server to meet specific requirements.
- **Cons**:
  - **Expensive**: Dedicated hosting is significantly costlier than shared hosting, with prices ranging from $100 to $500 per month or more.
  - **Requires Expertise**: Users need technical knowledge to manage and optimize the server.

- **Example Providers**:
  - Liquid Web, InMotion Hosting, and Rackspace.

3. Cloud Hosting

Cloud hosting leverages a network of interconnected virtual servers to host websites, providing unparalleled scalability and reliability.

- **How It Works**:
  - Website data is distributed across multiple servers in a cloud network, ensuring high availability and seamless resource allocation.
- **Ideal For**:
  - Websites experiencing fluctuating traffic or requiring dynamic scalability.
- **Pros**:
  - **Scalability**: Resources can be adjusted on demand, making it suitable for growing businesses.
  - **Reliability**: Data is stored across multiple servers, minimizing downtime.
  - **Cost-Efficiency**: Pay-as-you-go pricing models ensure that users only pay for the resources they consume.
- **Cons**:
  - **Complex Setup**: Requires technical knowledge to configure and manage.
  - **Variable Costs**: Costs can fluctuate based on resource usage, making budgeting challenging.
- **Example Providers**:
  - Amazon Web Services (AWS), Google Cloud Platform, and Microsoft Azure.

4. Virtual Private Server (VPS) Hosting

VPS hosting offers a middle ground between shared and dedicated hosting, providing virtualized server environments with dedicated resources.

- **How It Works**:
  - A physical server is divided into virtual partitions, each acting as an independent server for individual clients.
- **Ideal For**:
  - Medium-sized businesses and developers needing more control than shared hosting but without the expense of dedicated hosting.
- **Pros**:
  - **Greater Control**: Users have root access to their virtual environment, allowing for customization.

    - **Better Performance**: Dedicated resources ensure consistent speed and reliability.
    - **Cost-Effective**: Provides many of the benefits of dedicated hosting at a lower price.
- **Cons**:
    - **Requires Technical Knowledge**: Users must have some expertise to manage and optimize the VPS.
    - **Potential Resource Contention**: Overloaded servers may affect performance.
- **Example Providers**:
    - DigitalOcean, Linode, and Vultr.

5. Managed Hosting

Managed hosting is a fully managed solution where the hosting provider handles server maintenance, updates, backups, and security.

- **How It Works**:
    - Hosting providers take care of the technical aspects, allowing users to focus on their content and business.
- **Ideal For**:
    - Users with minimal technical expertise or those running platforms like WordPress.
- **Pros**:
    - **Reduced Workload**: Providers handle routine maintenance and updates.
    - **Enhanced Security**: Regular backups and proactive monitoring minimize risks.
    - **Time-Saving**: Allows users to focus on website content and business operations.
- **Cons**:
    - **Higher Costs**: Managed hosting plans are generally more expensive than unmanaged alternatives.
    - **Limited Customization**: Users may have less control over server configurations.
- **Example Providers**:
    - WP Engine, Kinsta, and Flywheel.

Key Features to Consider When Choosing a Hosting Service

Selecting the right hosting service is critical to the success and sustainability of any website or blog. Hosting providers vary widely in the features they offer, and understanding these features ensures that you choose a solution aligned with your technical needs, budget, and performance expectations. Below are the essential features to evaluate when selecting a hosting provider:

1. Uptime

Uptime refers to the percentage of time that a server is operational and accessible to users. A high uptime rate ensures that your website remains consistently available, minimizing the risk of downtime that could impact user experience and business operations.

- **Ideal Uptime Rate**:
    - Look for hosting providers offering a minimum of **99.9% uptime**, which translates to less than 8.77 hours of downtime per year.
    - Providers with "five nines" (99.999%) uptime guarantees are optimal for mission-critical websites, reducing downtime to just minutes annually.
- **Impact**:
    - Poor uptime affects user trust and search engine rankings, as consistent downtime signals unreliability to both users and algorithms.
- **Example**:
    - SiteGround is renowned for its high uptime rates, typically exceeding 99.9%.

2. Bandwidth

Bandwidth determines the amount of data that can be transferred between the website and its visitors over a given period. It directly impacts the number of visitors your site can handle simultaneously.

- **Factors to Consider**:
    - **Traffic Volume**: Websites expecting high traffic or hosting large media files require higher bandwidth.
    - **Unmetered Plans**: Some providers offer "unmetered" or "unlimited" bandwidth, though these often come with fair usage policies.
- **Impact**:
    - Insufficient bandwidth can lead to slow page loading times or downtime during traffic spikes, frustrating users.
- **Example**:
    - AWS provides scalable bandwidth options, making it suitable for websites with fluctuating traffic patterns.

3. Storage

Storage refers to the amount of disk space available to store website files, including text, images, videos, and databases. The storage requirements vary depending on the type of website:

- **Factors to Consider**:
    - **Basic Websites**: Blogs and small business sites may require as little as 10 GB.
    - **Media-Intensive Websites**: E-commerce, video streaming, and portfolio sites may need hundreds of gigabytes or even terabytes.

- **Types of Storage**:
  - **HDD (Hard Disk Drive)**: Cost-effective but slower and less durable.
  - **SSD (Solid State Drive)**: Faster, more reliable, and increasingly common in modern hosting plans.
- **Impact**:
  - Insufficient storage can limit website functionality, especially for sites with growing databases or frequent content updates.
- **Example**:
  - Bluehost offers SSD storage across its plans, ensuring fast and reliable performance.

4. Customer Support

Customer support is vital for resolving technical issues quickly and minimizing disruptions to your website. Hosting providers offer varying levels of support, including:

- **Support Channels**:
  - Live chat, phone, email, and support tickets are common options.
  - Look for providers offering 24/7 support for immediate assistance.
- **Additional Features**:
  - Knowledge bases, video tutorials, and community forums provide self-help resources.
- **Impact**:
  - Poor support can leave critical issues unresolved for extended periods, harming your website's performance and credibility.
- **Example**:
  - SiteGround is praised for its responsive and knowledgeable customer support, available via live chat and phone.

5. Scalability

Scalability ensures that your hosting plan can accommodate the growth of your website, whether in terms of traffic, content, or functionality.

- **Types of Scalability**:
  - **Vertical Scaling**: Upgrading resources like CPU, RAM, and storage within the same server.
  - **Horizontal Scaling**: Adding additional servers or resources to handle increased demand.
- **Impact**:
  - Scalability is crucial for websites experiencing seasonal spikes in traffic or rapid business growth.

- **Example**:
  - Cloud hosting providers like AWS and Google Cloud Platform offer dynamic scaling options to meet evolving needs.

6. Security

Security is a top priority for protecting websites from cyber threats, such as hacking, malware, and data breaches. Hosting providers often include a range of security features:

- **Key Security Features**:
  - **SSL Certificates**: Encrypt data transfer between the server and users, ensuring secure browsing.
  - **Firewalls**: Block unauthorized access and detect malicious activity.
  - **Regular Backups**: Protect against data loss due to hardware failure or cyberattacks.
- **Additional Measures**:
  - DDoS protection, spam filters, and two-factor authentication add extra layers of security.
- **Impact**:
  - A lack of robust security measures can lead to financial losses, reputational damage, and compromised user trust.
- **Example**:
  - SiteGround includes free SSL certificates, daily backups, and advanced firewall systems in its plans.

Example Hosting Services

To help you make an informed decision, here are some of the leading hosting providers and their standout features:

1. Bluehost

- **Specialization**: Known for WordPress hosting, ideal for beginners and small businesses.
- **Key Features**:
  - Free domain name for the first year.
  - One-click WordPress installation.
  - 24/7 customer support.

2. SiteGround

- **Specialization**: Robust security features and top-notch customer support.
- **Key Features**:
  - Managed WordPress hosting.

- Daily backups and advanced firewall systems.
- Fast performance with Google Cloud infrastructure.

3. AWS (Amazon Web Services)

- **Specialization**: Scalable cloud hosting for enterprise-level solutions.
- **Key Features**:
  - Pay-as-you-go pricing.
  - Elastic Load Balancing for handling traffic spikes.
  - High-level security and compliance.

4. HostGator

- **Specialization**: Affordable shared hosting with user-friendly tools.
- **Key Features**:
  - Unmetered bandwidth.
  - Free website builder.
  - 24/7 support via phone and live chat.

5. WP Engine

- **Specialization**: Managed hosting for WordPress sites.
- **Key Features**:
  - Automatic updates and backups.
  - Advanced performance monitoring tools.
  - 24/7 expert WordPress support.

2. Introduction to HTML

HTML (HyperText Markup Language) is the cornerstone of web development, providing the framework for creating, structuring, and displaying content on the internet. It defines how text, images, links, multimedia, and interactive elements are organized and rendered on a webpage. As the foundational language of the web, HTML is indispensable for web designers, developers, and anyone aiming to establish an online presence.

First introduced by Tim Berners-Lee in 1991, HTML has undergone significant advancements, with the current standard being **HTML5**. Each iteration has enhanced its capabilities, making it more robust, versatile, and compatible with modern web technologies. From simple static websites to complex dynamic platforms, HTML plays a central role in ensuring accessibility, usability, and interactivity. At its core, HTML serves as the skeleton of a webpage. It uses a system of elements and tags to specify the structure and layout of content. These elements define everything from headings, paragraphs, and lists to images, videos, and links. However, HTML is not limited to static content. It integrates seamlessly with **CSS (Cascading Style Sheets)** for styling and **JavaScript** for interactivity, enabling the creation of visually appealing and functionally rich websites.

Key Features of HTML

HTML is celebrated for its simplicity, flexibility, and adaptability. Its features make it a universal standard for web development, enabling creators to build websites that cater to diverse audiences and platforms.

1. Tag-Based Structure

HTML employs a tag-based syntax to define and organize content. Tags are enclosed in angle brackets (<>) and often come in pairs: an opening tag (<p>) and a closing tag (</p>), with the content placed in between.

- **Examples**:
  - <h1>: Defines a primary heading.
  - <p>: Represents a paragraph.
  - <a>: Creates hyperlinks.

Each tag serves a specific purpose, allowing developers to structure content hierarchically and logically.

2. Platform Independence

One of HTML's greatest strengths is its platform independence. HTML files can be rendered on any device or browser, making it a universal language for the web. Whether viewed on a smartphone, tablet, or desktop, properly coded HTML ensures that the content remains accessible and functional.

- **Cross-Browser Compatibility**:
  - HTML5 is supported by all major browsers, including Chrome, Firefox, Safari, and Edge.

3. Scalability

HTML's scalability enables it to accommodate projects of varying complexity. From simple single-page websites to large-scale e-commerce platforms, HTML adapts to the needs of the project. Its modular nature allows developers to build upon existing structures, scaling up as required.

4. Integration with Other Technologies

HTML works seamlessly with other web technologies to enhance functionality and design:

- **CSS**:
  - Controls the visual presentation of HTML elements, including layout, colors, fonts, and animations.
- **JavaScript**:
  - Adds interactivity to web pages, such as dynamic forms, slideshows, and interactive maps.

This integration enables the creation of sophisticated, user-friendly websites that cater to modern demands.

Basic HTML Tags

HTML tags are the building blocks of a webpage, defining its structure and content. Understanding these tags is fundamental for creating effective and organized web pages.

1. Structural Tags

Structural tags define the core framework of an HTML document. They ensure proper organization and compatibility with browsers.

- **<html>**:
    - Represents the root element of the HTML document.
    - Encloses all other elements.
- **<head>**:
    - Contains metadata about the document, such as the title, character encoding, and links to stylesheets or scripts.
    - Example:

```
<head>
 <title>My Web Page</title>
</head>
```

- **<body>**:
    - Encloses the content visible to users, such as text, images, and links.
    - Example:

```
<body>
 <h1>Welcome to My Website</h1>
 <p>This is my first paragraph.</p>
</body>
```

2. Text Formatting Tags

Text formatting tags control how textual content appears on the webpage. They define headings, paragraphs, emphasis, and other text-related styles.

- **Headings (<h1> to <h6>)**:
    - Define headings of varying importance, with <h1> being the most prominent and <h6> the least.
    - Example:

```
<h1>Main Heading</h1>
```

```
<h2>Subheading</h2>
```

- **Paragraph (<p>):**
  - Group related text into a block.
  - Example:

```
<p>This is a paragraph of text.</p>
```

- **Bold and Italics (<b> and <i>):**
  - Add emphasis to text.
  - Example:

```
<b>Bold Text</b> and <i>Italic Text</i>
```

3. Link and Media Tags

Links and media tags are essential for adding interactivity and multimedia elements to web pages.

- **Hyperlinks (<a>):**
  - Creates clickable links to internal or external resources.
  - Example:

```
<a href=”https://example.com”>Visit Example</a>
```

- **Images (<img>):**
  - Embed images into the page.
  - Attributes include src (source URL) and alt (alternative text for accessibility).
  - Example:

```
<img src=”image jpg” alt-”A beautiful sunset”>
```

4. List Tags

Lists organize content into bullet points or numbered sequences, improving readability and presentation.

- **Unordered List (<ul>):**
  - Displays items as bullet points.
  - Example:

```
<ul>
  <li>Item 1</li>
  <li>Item 2</li>
</ul>
```

- **Ordered List (<ol>):**
  - Displays items in a numbered sequence.

- Example:

```
<ol>
  <li>First</li>
  <li>Second</li>
</ol>
```

HTML Document Structure

An HTML document follows a standardized structure to ensure that browsers interpret and render it correctly.

HTML Boilerplate Example

```
<!DOCTYPE html>
<html>
<head>
  <title>Sample Page</title>
</head>
<body>
  <h1>Welcome to HTML</h1>
  <p>This is a paragraph in a well-structured HTML document.</p>
</body>
</html>
```

Explanation of Key Components

1. **<!DOCTYPE html>**:
   - Declares the document type as HTML5.
2. **<html>**:
   - The root element that encloses all content.
3. **<head>**:
   - Contains metadata, such as the page title and linked resources (e.g., stylesheets).
4. **<body>**:
   - Houses the visible content, including text, images, and multimedia.

Why HTML is Essential

1. **Accessibility**:
   - Proper HTML ensures that websites are accessible to users with disabilities, enhancing inclusivity.

2. **SEO**:
    - Clean, semantic HTML helps search engines index and rank content effectively.
3. **Interoperability**:
    - HTML's compatibility with all browsers and platforms ensures a consistent user experience.
4. **Foundation for Advanced Development**:
    - HTML serves as the foundation for integrating advanced features through CSS, JavaScript, and frameworks like React or Angular.

3. HTML Document Structure

The structure of an HTML document is fundamental to its functionality, ensuring that web browsers can interpret and render the content as intended. A well-organized HTML document adheres to a standard format that promotes compatibility, accessibility, and usability. This structure is like the blueprint of a building—it defines the framework within which all components are placed, from metadata to visible elements like text, images, and links. An HTML document is essentially a hierarchical arrangement of elements, with each part serving a specific role in the overall functionality and presentation of the webpage. The structure not only facilitates browser rendering but also enhances user experience and search engine optimization (SEO), making it a critical aspect of web design and development.

HTML Boilerplate

The following example represents the basic structure of an HTML document, often referred to as a "boilerplate." This template forms the foundation of any HTML page.

```
<!DOCTYPE html>
<html>
  <head>
    <title>My First Web Page</title>
    <meta charset="UTF-8">
    <meta name="viewport" content="width=device-width, initial-scale=1.0">
  </head>
  <body>
    <h1>Welcome to My Website</h1>
    <p>This is a paragraph.</p>
  </body>
</html>
```

Explanation of the Structure

Each part of the HTML document serves a unique function, contributing to the page's functionality, appearance, and compatibility.

1. <!DOCTYPE html> Declaration

- The DOCTYPE declaration specifies the version of HTML being used, which is HTML5 in this case. It helps browsers render the content correctly by informing them of the document's syntax rules.
- **Example**:

  <!DOCTYPE html>
- **Purpose**:
  - Ensures consistency in rendering across different browsers.
  - Helps prevent quirks mode, where browsers may interpret the document inconsistently.

2. Root Element (<html>)

- The <html> tag is the root element that encloses the entire HTML document. It serves as the container for all other elements.
- **Attributes**:
  - lang: Specifies the language of the document, aiding in accessibility and SEO.

  <html lang="en">

Purpose:

  - Organizes the document's content.
  - Enhances accessibility by providing language information to screen readers and search engines.

3. Head Section (<head>)

The <head> section contains metadata and resources that are not directly visible on the webpage but are essential for its operation and optimization.

- **Key Elements in the <head>**:
  1. **<title>**:
     - Sets the title displayed on the browser tab or search engine results.
     - Example:

  <title>My First Web Page</title>

  2. **<meta>**:
     - Provides metadata about the webpage.

- Example:

<meta charset="UTF-8">

- Specifies the character encoding, ensuring proper display of text.

<meta name="viewport" content="width=device-width, initial-scale=1.0">

- Makes the page responsive by adjusting its layout based on the device's screen size.

3. **External Resources:**
   - Links to stylesheets, JavaScript files, or icons.
   - Example:

<link rel="stylesheet" href="styles.css">

4. Body Section (<body>)

The <body> section contains all the content visible to users, including text, images, links, videos, and interactive elements. It is where the primary focus of design and development lies.

- **Examples of Common Elements in <body>:**
  1. **Headings:**
     - Define hierarchical sections of content.
     - Example:

<h1>Welcome to My Website</h1>

  2. **Paragraphs:**
     - Enclose blocks of text.
     - Example:

<p>This is a sample paragraph.</p>

  3. **Links:**
     - Provide navigation to other pages or resources.
     - Example:

<a href="https://example.com">Visit Example</a>

  4. **Images:**
     - Embed visual content.
     - Example:

<img src="image.jpg" alt="Sample Image">

Importance of Proper Structure

A properly structured HTML document is essential for various technical and functional reasons. It enhances the page's performance, usability, and compliance with web standards.

1. Browser Compatibility

- A well-structured HTML document ensures that all browsers can interpret and display the content consistently.
- Browsers rely on the standardized structure to parse the document efficiently.
- Example:
    - If the <meta charset="UTF-8"> tag is missing, some special characters may not render correctly.

2. Accessibility

- Clear organization aids assistive technologies like screen readers in navigating the webpage.
- Tags like <h1> and <p> provide semantic meaning, helping users with disabilities understand the content.
- Example:
    - A <nav> tag helps screen readers identify navigation menus.

3. Search Engine Optimization (SEO)

- Search engines prioritize well-structured HTML documents for indexing and ranking.
- Semantic tags, such as <header>, <main>, and <footer>, improve the SEO value of a page.
- Example:
    - A <title> tag with relevant keywords boosts visibility in search results.

Common Mistakes to Avoid

1. **Omitting the DOCTYPE Declaration**:
    - Leads to quirks mode, causing inconsistent rendering.
2. **Improper Nesting of Tags**:
    - Tags must be closed in the reverse order of opening.
    - Incorrect:

```
<p><strong>Bold Text</p></strong>
```

    - Correct:

```
<p><strong>Bold Text</strong></p>
```

3. **Missing Metadata**:
    - Failing to include viewport settings can result in poor mobile responsiveness.

4. **Empty Tags**:
    - Ensure that tags like <img> always have attributes like src and alt.

Practical Applications of HTML Document Structure

1. **Building Responsive Websites**:
    - A well-structured document with proper metadata ensures that the page adapts to various screen sizes.
2. **Creating Accessible Content**:
    - Using semantic tags improves navigation for users relying on assistive technologies.
3. **Enhancing Collaboration**:
    - A standardized structure makes it easier for teams to work together, as the code is organized and understandable.

Chapter 8

Cascading Style Sheets (CSS)

Cascading Style Sheets (CSS) is a cornerstone of web design, enabling developers to control the presentation and layout of HTML content. While HTML provides the structure of a webpage, CSS adds the style, transforming plain text and simple layouts into visually engaging and user-friendly designs. Introduced in 1996, CSS has become an essential tool for separating content from design, allowing for cleaner, more maintainable code. CSS empowers developers to control a wide range of stylistic properties, including colors, fonts, margins, borders, and positioning. By using CSS, designers can create consistent and responsive designs that work seamlessly across devices and browsers. This flexibility makes CSS an indispensable skill for anyone involved in web development.

1. Core Concepts of CSS

Cascading Style Sheets (CSS) revolutionize web design by providing a structured and efficient way to apply visual styling to HTML content. It simplifies the process of managing the appearance of web pages, allowing for a centralized control of design rules that can be applied consistently across multiple elements, pages, or even entire websites. CSS empowers developers and designers to create visually appealing, responsive, and user-friendly websites without compromising on maintainability or performance.

Separation of Content and Presentation

One of the fundamental principles of CSS is the separation of content (HTML) from presentation (CSS). This approach enhances code clarity, maintainability, and scalability while ensuring that design changes can be implemented quickly without altering the core content structure.

- **HTML for Structure**:
  - HTML defines the basic layout and organization of content on a webpage.
  - Example:

```
<h1>Welcome to My Website</h1>
<p>This is an introductory paragraph.</p>
```

- **CSS for Styling**:
  - CSS applies the visual presentation, such as colors, fonts, and layout, to the structured HTML.

DOI: 10.4324/9781003750444-8

- Example:

```
h1 {color: blue;
 font-size: 24px; font-family: Arial, sans-serif;}
p { color: gray;  line-height: 1.5;}
```

- **Benefits**:
  - **Reusability**: CSS rules can be applied to multiple pages or elements.
  - **Consistency**: Ensures a uniform appearance across a website.
  - **Ease of Updates**: Design changes require modifying only the CSS file, leaving HTML untouched.

The Cascading Nature of CSS

The term "cascading" in CSS refers to the hierarchy and prioritization of style rules when multiple declarations target the same element. CSS follows a well-defined set of rules to determine which style is applied, ensuring flexibility and control for designers.

Hierarchy of Rules

CSS prioritizes styles based on their source and specificity. The order of precedence is as follows:

1. **Inline Styles**:
   - Defined directly within an HTML element using the style attribute.
   - **Example**:

```
<h1 style="color: red; font-size: 28px;">Inline Styled Heading</h1>
```

   - **Advantages**:
     - Useful for testing or applying quick, element-specific styles.
   - **Disadvantages**:
     - Mixes content and presentation, making code harder to maintain.
2. **Internal Stylesheets**:
   - Defined within the <style> tag in the <head> section of an HTML document.
   - **Example**:

```
h1 {color: green; }
```

   - **Advantages**:
     - Centralized styling for a single HTML document.
   - **Disadvantages**:
     - Not reusable across multiple documents.

3. **External Stylesheets**:
    - Stored in separate .css files and linked to the HTML document using the <link> tag.
    - **Example**:

```
<link rel="stylesheet" href="styles.css">
```

- styles.css:

```
h1 { color: blue; font-size: 24px;}
```

- **Advantages**:
    - Promotes reusability and consistency across multiple pages.
    - Simplifies maintenance and collaboration.
- **Disadvantages**:
    - Requires an additional HTTP request to load the stylesheet.

Specificity

CSS specificity determines which rule is applied when multiple rules target the same element. It is calculated based on the type of selector used:

1. **Inline Styles**: Have the highest specificity.
    - Example: <h1 style="color: red;">Inline Style</h1>.
2. **ID Selectors**: More specific than class or type selectors.
    - Example:

```
#main-heading { color: blue;}
```

3. **Class Selectors**: More specific than type selectors.
    - Example:

```
.highlight { color: green;}
```

4. **Type Selectors**: Have the lowest specificity.
    - Example:

```
h1 { color: black;}
```

- **Example of Conflicting Rules**:

```
<style>
 h1 { color: black;} .highlight { color: green;} #main-heading { color: blue; }
</style>
<h1 id="main-heading" class="highlight">Styled Heading</h1>
```

- The final color applied to the heading is **blue** because the ID selector has the highest specificity.

Inheritance

Certain CSS properties, such as color and font-family, are inherited from parent elements to child elements, simplifying the styling process.

- **Example**:

```
<style>
 body {color: darkgray;
  font-family: Arial, sans-serif; }
</style>
<body>
 <h1>This heading inherits the color and font-family.</h1>
 <p>So does this paragraph.</p>
</body>
```

- **Non-Inherited Properties**:
    - Properties like margin and padding must be explicitly defined.

The Importance of Cascading and Specificity

The cascading nature of CSS ensures flexibility in styling while maintaining control over which rules are applied. It allows developers to:

- Override default styles when necessary.
- Apply global styles across an entire website using low-specificity selectors.
- Target specific elements with high-specificity selectors or inline styles.

Example Use Case: Applying the Core Concepts

Consider a simple webpage styled using all three methods—inline, internal, and external CSS:

- **HTML Document**:

```
<!DOCTYPE html>
<html lang="en">
 <head>
  <title>My Styled Page</title>
  <style>
   body {font-family: Arial, sans-serif; }
   h1 { color: green; }
  </style>
  <link rel="stylesheet" href="styles.css">
```

```
  </head>
  <body>
    <h1 style="color: red;">Welcome to My Website</h1>
    <p>This is a paragraph styled using an external stylesheet.</p>
  </body>
</html>
```

- **External Stylesheet (styles.css)**:

```
p { color: blue; line-height: 1.5;}
```

- **Result**:
    - The <h1> element appears in **red** because the inline style has the highest specificity.
    - The <p> element appears in **blue** with a line height of 1.5, as styled by the external stylesheet.

2. Types of CSS

CSS (Cascading Style Sheets) can be implemented in three main ways: **Inline CSS, Internal CSS**, and **External CSS**. Each method serves specific purposes and offers distinct advantages and disadvantages. Understanding these types is crucial for determining the most effective approach for styling a webpage based on its requirements, scale, and complexity.

a. Inline CSS

Inline CSS involves defining styles directly within the HTML elements using the style attribute. This method is suitable for applying unique styles to a specific element or for quick testing during development.

- **Syntax Example**:

```
<h1 style="color: blue; font-size: 24px;">Welcome to My Website</h1>
```

Features:

- Inline styles are applied immediately to the element in question, without affecting other elements.

Advantages:

1. **Quick and Simple**:
    - Ideal for small changes or testing during development.
2. **No External Dependencies**:
    - Does not require linking to external stylesheets or defining rules elsewhere.
3. **Specificity**:
    - Inline styles have the highest priority among CSS rules, making them useful for overriding other styles.

Disadvantages:

1. **Poor Maintainability**:
   - Mixing content (HTML) and presentation (CSS) complicates code readability and maintainability.
2. **Repetition**:
   - Applying the same styles to multiple elements requires repeating the style attribute, increasing redundancy.
3. **Not Scalable**:
   - Inefficient for large projects or pages with extensive styling needs.

- **Use Case**:
  - Inline CSS is best suited for small tweaks or one-off styles that do not require reuse.

b. Internal CSS

Internal CSS is defined within the <style> tag in the <head> section of an HTML document. This method allows for centralized styling of a single page, making it more organized compared to inline CSS.

- **Syntax Example**:

```
h1 {color: green; font-family: Arial, sans-serif;}
p { color: gray; line-height: 1.5;}
```

Features:

- Styles defined in the <style> tag are scoped to the current HTML document, allowing for consistent styling across the page.

Advantages:

1. **Organized Code**:
   - Centralizes all styles for a single page, making it easier to manage compared to inline CSS.
2. **Better Readability**:
   - Keeps the HTML elements clean by separating content from styling.
3. **No Additional Files**:
   - Eliminates the need for linking external stylesheets.

Disadvantages:

1. **Limited Reusability**:
   - Styles cannot be reused across multiple pages, making it inefficient for multi-page websites.

2. **Page-Specific**:
   - Changes must be made in each HTML file if multiple pages require the same styles.

- **Use Case**:
  - Internal CSS is suitable for single-page websites, landing pages, or projects where consistency is needed but reuse across multiple pages is not required.

c. External CSS

External CSS involves storing styles in a separate .css file and linking it to the HTML document using the <link> tag. This is the most efficient and scalable method for styling websites.

- **Syntax Example**: **HTML**:

<link rel="stylesheet" href="styles.css">

styles.css:

h1 { color: red; text-align: center}

p {font-size: 18px; line-height: 1.}

Features:

- External stylesheets can be reused across multiple HTML pages, ensuring consistency and reducing redundancy.

Advantages:

1. **Reusability**:
   - A single CSS file can style an entire website, promoting consistency.
2. **Scalability**:
   - Suitable for projects of any size, from small blogs to large e-commerce platforms.
3. **Separation of Concerns**:
   - Keeps HTML files clean by separating content from styling.
4. **Maintainability**:
   - Updates to the CSS file instantly reflect across all linked pages, reducing workload.

Disadvantages:

1. **Dependency on External Files**:
   - The webpage relies on the external CSS file; if the file fails to load, the page may lose its styling.

2. **Additional HTTP Requests**:
   - Linking to an external stylesheet increases the number of HTTP requests, potentially affecting load times.

- **Use Case**:
  - External CSS is the standard choice for multi-page websites or projects requiring consistent and scalable design systems.

Comparison of CSS Types

Aspect	**Inline CSS**	**Internal CSS**	**External CSS**
Definition	Styles written directly within an HTML element.	Styles defined within the <style> tag in <head>.	Styles stored in a separate .css file.
Reusability	None	Limited to a single page.	High, reusable across multiple pages.
Code Organization	Poor, as it mixes content and styling.	Better, but still limited to one page.	Excellent, separates content from styling.
Maintainability	Low	Moderate	High
Scalability	Not scalable.	Suitable for single-page websites.	Highly scalable for large projects.
Use Case	Quick fixes or testing.	Single-page styling.	Multi-page projects requiring consistent styling.

Choosing the Right CSS Type

Selecting the appropriate type of CSS depends on the project's requirements:

- **Use Inline CSS** for:
  - Small, one-off changes or testing purposes.
  - Adding styles that will not be reused.
- **Use Internal CSS** for:
  - Single-page applications where centralized styling is sufficient.
  - Projects with unique page-specific styling requirements.
- **Use External CSS** for:
  - Large-scale or multi-page websites that require consistent design.
  - Projects where maintainability and scalability are priorities.

Practical Example: Combining CSS Types

In some cases, a combination of CSS types may be used to achieve specific goals. For example:

```
<!DOCTYPE html>
```

```
<html lang="en">
  <head>
    <title>Combining CSS Types</title>
    <style>
      /* Internal CSS */
      h1 {color: green;
        font-size: 36px;}
    </style>
    <link rel="stylesheet" href="styles.css">
  </head>
  <body>
    <h1 style="text-align: center;">Welcome</h1> <!-- Inline CSS -->
    <p>This paragraph is styled using external CSS.</p>
  </body>
</html>
```

3. CSS Syntax

CSS syntax consists of selectors, properties, and values, forming the foundation of every CSS rule.

Basic Structure

```
selector {property: value;}
```

Components

1. **Selector**:
    - Specifies the HTML element(s) to which the style applies.
    - Example: h1, .class-name, or #id-name.
2. **Property**:
    - Defines the aspect of the element to style, such as color or margin.
3. **Value**:
    - Specifies the desired style for the property.
    - Example: blue, 20px.

- **Example Rule**:

```
p {color: black;line-height: 1.5;}
```

4. Key Features of CSS

Selectors

CSS selectors target specific elements in an HTML document. Common types include:

1. **Type Selectors**:
   - Target HTML tags.
   - Example:

h1 {font-size: 28px;}

2. **Class Selectors**:
   - Target elements with a specific class attribute.
   - Example: .highlight {background-color: yellow;}
3. **ID Selectors**:
   - Target elements with a unique ID attribute.
   - Example: #main-title {color: purple;}
4. **Group Selectors**:
   - Apply the same styles to multiple elements.
   - Example:

h1, h2, p {font-family: Arial, sans-serif;}

Box Model

The CSS box model is a fundamental concept that describes how elements are structured on a webpage.

1. **Content**:
   - The actual content inside the element.
2. **Padding**:
   - Space between the content and the element's border.
3. **Border**:
   - Surrounds the padding.
4. **Margin**:
   - Space between the element and adjacent elements.

- **Example**:

div {width: 200px; padding: 10px; border: 2px solid black; margin: 20px;}

Responsive Design

CSS facilitates responsive web design, ensuring that websites adapt to different screen sizes and devices.

1. **Media Queries**:
    - Apply styles based on device characteristics, such as screen width.
    - Example:

@media (max-width: 768px) {body {background-color: lightgray;}}

2. **Flexible Layouts**:
    - Use relative units like percentages or em to create fluid layouts.

5. Advanced CSS Techniques

CSS has evolved to include advanced features, enabling modern web design:

1. **Transitions**:
    - Add smooth effects to property changes.
    - Example:

button {transition: background-color 0.3s ease;}

2. **Animations**:
    - Create dynamic effects.
    - Example: @keyframes fadeIn {from {opacity: 0;} to {opacity: 1;}}

div {animation: fadeIn 2s;}

3. **Flexbox and Grid**:
    - Simplify complex layouts.
    - **Flexbox** Example:.container {display: flex; justify-content: space-between;}

6. Applications of CSS

CSS is widely used in web development to enhance the visual appeal and functionality of websites:

1. **Theming**:
    - Define consistent color schemes and fonts.
    - Example: Dark and light mode styles.
2. **Interactive Interfaces**:
    - Style buttons, hover effects, and forms to improve user engagement.
3. **Responsive Design**:
    - Ensure optimal viewing experiences across devices.

Text Styling in CSS and Its Functionality

Text styling in CSS is a fundamental aspect of web design, enhancing not only the visual appeal but also the usability and accessibility of digital content. Properly styled text ensures that information is engaging, legible, and accessible across various devices, platforms, and user needs. CSS offers an extensive set of properties to control every detail of text presentation, from fonts and alignment to spacing, color, and responsiveness.

Key Text Properties in CSS

CSS provides a robust toolkit of text-related properties, enabling developers to design content that is both functional and visually captivating. These properties are categorized based on their impact on the text's appearance, spacing, and alignment.

1. Font Properties

Font properties define the typeface, size, weight, style, and other characteristics of the text, allowing designers to create a specific visual tone or branding identity.

Font Family (font-family)

- Specifies the typeface used for the text.
- Developers can choose from predefined font families or integrate custom fonts using online services like Google Fonts.
- **Fallback Mechanism**:
    - When a browser does not support the primary font, additional fonts listed as fallbacks ensure consistent rendering.

Font Size (font-size)

- Determines the size of the text.
- Can be defined using various units, including:
    - **Pixels (px)**: Fixed units suitable for precise control.
    - **Ems (em)**: Relative to the parent element's font size.
    - **Percentages (%)**: Scales based on the parent element.
    - **Keywords**: Values like small, medium, or large.

Font Weight (font-weight)

- Specifies the thickness of the font.
- Accepts named values such as normal and bold or numeric values ranging from 100 (thin) to 900 (extra bold).

Font Style (font-style)

- Applies stylistic variations to the text, such as italicization or oblique rendering.

- Common values include normal, italic, and oblique.

Font Variant (font-variant)

- Enables stylistic effects like small caps.
- Commonly used for headings or decorative text elements.

2. Text Alignment and Spacing

Proper alignment and spacing are crucial for ensuring readability and creating aesthetically pleasing layouts.

Text Alignment (text-align)

- Aligns text horizontally within its container.
- Common values:
  - left: Aligns text to the left (default for most languages).
  - right: Aligns text to the right.
  - center: Centers the text.
  - justify: Aligns text to both margins, creating a clean block-like appearance.

Line Height (line-height)

- Adjusts the vertical spacing between lines of text.
- Ideal for improving readability, especially in paragraphs with dense information.

Letter Spacing (letter-spacing)

- Modifies the spacing between individual letters.
- Used for creating distinct visual effects or improving readability in decorative fonts.

Word Spacing (word-spacing)

- Adjusts the distance between words.
- Useful for creating visually balanced text layouts.

Text Indentation (text-indent)

- Indents the first line of a paragraph.
- Commonly used in essays, articles, or long-form content.

3. Text Decoration

Text decoration properties define additional visual effects applied to text, such as underlining, striking through, or capitalizing.

Text Decoration (text-decoration)

- Applies or removes visual enhancements like underlines.
- Common values include none, underline, overline, and line-through.

Text Transform (text-transform)

- Controls capitalization of text.
- Values include:
    - uppercase: Converts text to all capital letters.
    - lowercase: Converts text to all lowercase letters.
    - capitalize: Capitalizes the first letter of each word.

4. Text Color

Color (color)

- Sets the color of the text.
- Acceptable formats include:
    - **Named Colors**: e.g., red, blue, green.
    - **Hexadecimal Codes**: e.g., #333333, #FF5733.
    - **RGB Values**: e.g., rgb(51, 51, 51).
    - **HSL Values**: e.g., hsl(0, 0%, 20%).

Working with Text in CSS

Beyond basic styling, CSS empowers developers to create responsive, accessible, and interactive text designs. These capabilities ensure that text is not only visually appealing but also functional for a diverse range of users and devices.

1. Responsive Text

Responsive design ensures that text remains legible and aesthetically pleasing across various screen sizes and devices.

Relative Units

- Using relative units like em or percentages allows text size to scale dynamically with the viewport or parent element.

Media Queries

- Media queries adjust text styles based on specific screen characteristics such as width, height, or resolution.
- Example:

- Larger font sizes for desktop displays.
- Smaller, more compact fonts for mobile screens.

2. Accessibility in Text Styling

Accessible text styling is essential for creating inclusive web experiences, particularly for users with visual impairments or cognitive challenges.

Contrast

- Ensure sufficient contrast between text and background colors to enhance readability.

Scalable Fonts

- Allow users to adjust font sizes easily by using relative units like em or %.

Semantic HTML

- Use proper HTML elements (<h1> to <h6>, <p>, <blockquote>) to provide structural meaning and facilitate navigation for screen readers.

Best Practices for Text Styling in CSS

1. **Consistency**:
   - Use a consistent font family and size hierarchy throughout the website to maintain a cohesive look and feel.
2. **Readability**:
   - Prioritize readability by choosing appropriate font sizes, line heights, and contrast levels.
3. **Minimal Use of Inline Styles**:
   - Avoid inline styles to maintain clean and maintainable code.
4. **Responsive Design**:
   - Use relative units and media queries to ensure text adapts to various screen sizes.
5. **Testing**:
   - Test text styling on multiple devices and browsers to ensure compatibility.

References

Andres D and Woodard J. 2013. *Social media handbook for agricultural development practitioners.* Publication by FHI360 of USAID. http://ictforag.org/toolkits/ social/ SocialMedia4 AgHandbook.pdf

Barber J, Mangnus E and Bitzer V. 2016. *Harnessing ICT for agricultural extension.* KIT Working Paper 2016: 4. https://213ou636sh0ptphd141fqei1-wpengine.netdna-ssl.com/sed/wp-content/uploads/sites/ 2/2016/11/KIT_WP2016-4_Harnessing-ICT-for-agricultural-extension.pdf

Bheenick K and Bionyi I. 2017. *Effective Tools for Knowledge Management and Learning in Agriculture and Rural Development.* CTA Working paper. https://publications.cta.int/media/publications/downloads/1986_PDF.pdf

Fafchamps M and Minten B. 2012. *Impact of SMS based Agricultural Information on Indian Farmers.* The World Bank Economic Review, Published by the Oxford University Press on behalf of the International Bank for Reconstruction and Development.

FAO 2011. *E-learning methodologies a guide for designing and developing e-learning courses.* Food and Agriculture Organization of the United Nations.http://www.fao.org/docrep/015/i2516e/i2516e.pdf

George T, Bagazonzya H, BallantyneP, Belden C, Birner R, Del CR and Treinen S. 2017. *ICT in agriculture: connecting smallholders to knowledge, networks, and institutions.* Washington,

Heike Baumüller. 2018. The little we know: An exploratory literature review on the utility of mobile phone enabled services for smallholder farmers. *Journal of International Development.* 30, 134–154.

Laurens K. 2016. *NELK Module 6: Basic Knowledge Management and Extension*, New Extensionist Learning Kit (NELK), Global Forum for Rural Advisory Services (GFRAS). http://www.g-fras.org/en/knowledge/new-extensionist-learning-kit-nelk.html# module-6-basic-knowledge-management-and-extension

Mayer RE. 2005. *The Cambridge handbook of multimedia learning.* New York: University of Cambridge. MEAS & Access Agriculture 2013. *A Guide to Producing Farmer-to-Farmer Training Videos.* https://www.agrilinks.org/sites/default/files/resource/files/MEAS%20Guide%20to% 20Producing%20 Farmer-to-Farmer%20Training%20 Videos%202013_04.pdf

Meera SN.2013. *Extension, ICTs and Knowledge Management: The 10 difficult questions.* Blog 15. Agricultural Extension in South Asia. http://www.aesanetwork.org/extension-icts-and-knowledge-management-the-10-difficultquestions/

Meera SN. 2017. *Disruptive Technologies – Big Data and Internet of Things in Strengthening Extension & Advisory Services.* Blog 68. Agricultural Extension in South Asia. http://www.aesanetwork.org/disruptive-technologies-big-data-and-internet-of-things-instrengthening- extension-advisory-services/

Meera SN. 2018. *A Treatise on Navigating Extension and Advisory Services through Digital Disruption.* Blog 90. Agricultural Extension in South Asia. http://www.aesanetwork.org/atreatise- on-navigating-extension-and-advisory-services-through-digital-disruption/

Mittal N, Surabhi, Gandhi, Sanjay and Gaurav T. 2010. *Socio-Economic Impact of Mobile Phones on Indian Agriculture.* ICRIER Working Paper No. 246, Indian Council for Research on International Economic Relations (ICRIER), New Delhi.

Preece J, Rogers Y, & Preece, J. 2007. *Interaction design: Beyond human-computer interaction.* Chichester: Wiley. Saravanan R, Sulaiman RV, Davis K and Suchiradipta B. 2015. *Navigating ICTs for Extension and Advisory Services. Note 11. GFRAS Good Practice Notes for Extension and Advisory Services.* GFRAS: Lindau, Switzerland. https://agrilinks.org/sites/default/files/resource/files/gfras-ggp-note11_navigating_icts_ for_ras_1.pdf

Saravanan R and Suchiradipta B. 2015. *mExtension – Mobile Phones for Agricultural Advisory Services.* Note 17. GFRAS Good Practice Notes for Extension and Advisory Services. GFRAS: Lindau, Switzerland. www.g-fras.org/en/download.html?download=349: ggp-note-17-mextension-mobile-phonesfor- agricultural-advisory-services

Saravanan R and Suchiradipta B. 2016. *Social media policy guidelines for agricultural extension and advisory services,* GFRAS interest group on ICT4RAS, GFRAS: Lindau, Switzerland. www.g-fras.org/en/knowledge/gfras-publications.html?download =415: social-media-policyguidelines- for-agricultural-extension-and-advisory-services

Saravanan R. 2010. (Ed.) *ICTs for Agricultural Extension: Global Experiments, Innovations and Experiences,* New India Publishing Agency (NIPA), New Delhi. http://www.saravananraj.net/wp-content/uploads/2014/12/32_India_ICTs-for-Agricultural- Extension_Saravanan.pdf

Saravanan R, Suchiradipta B, Chowdhury A, Hambly OH and Hall K. 2015. *Social Media for Rural Advisory Services.* Note 15. GFRAS Good Practice Notes for Extension and Advisory Services. GFRAS: Lindau, Switzerland. www.g-fras.org/en/download.html?download=355: ggp-note-15-social-media-for-rural advisory- services

Saravanan R, Suchiradipta B, Meera SN, Kathiresan C and Anandaraja N. 2015. *Web Portals for Agricultural Extension and Advisory Services.* Note 16. GFRAS Good Practice Notes for Extension and Advisory Services. GFRAS: Lindau, Switzerland. www.gfras.org/en/download.html?download=356: gfras-ggp-note-16-web-portals-for agricultural- extension-and-advisory-services

Saravanan R.2014. (Ed.). *Mobile Phones for Agricultural Extension: Worldwide mAgri Innovations and Promise for Future,* New India Publishing Agency, New Delhi. http://www.saravananraj.net/wp-content/uploads/2014/12/27_Mobile-phones-for- Agricultural-Extension-in-India_Saravanan-Raj-Draft.pdf

Saravanan R, Kathiresan C, and Indra DT. 2011. (Eds.) *Information and Communication Technology for Agriculture and Rural Development,* New India Publishing Agency (NIPA), New Delhi.

Sophie T and Alice VDE.2018. *Gender and ICTs - Mainstreaming gender in the use of information and communication technologies (ICTs) for agriculture and rural development,* FAO. http://www.fao.org/publications/card/en/c/I8670EN

Suchiradipta B and Saravanan R. 2016. *Social media: Shaping the future of agricultural extension and advisory services,* GFRAS interest group on ICT4RAS discussion paper, GFRAS: Lindau, Switzerland. www.g-fras.org/en/knowledge/gfras-publications.html? download=414: social-mediashaping- the-future-of-agricultural-extension-and-advisory-services

Vignare K. 2013. *Options and strategies for information and communication technologies within agricultural extension and advisory services.* MEAS Discussion paper. http://meas.illinois.edu/wpcontent/uploads/2015/04/Vignare-K-2013-ICT-and-Extension- MEAS-Discussion-Paper.pdf

World Bank. 2017. *ICT in Agriculture (Updated Edition): Connecting Smallholders to Knowledge, Networks, and Institutions.* Washington, DC: World Bank. https://openknowledge.worldbank.org/handle/10986/27526

Andleig PK and Thakrar K. 2003. Multimedia Systems Design. PHI: New Delhi.

Chrisanthi Avgerou, Robin Mansell, Danny Quah and Roger Silverstone. 2009. The Oxford Handbook of Information and Communication Technologies. http://www.oxfordhandbooks.com/view/10.1093/oxfordhb/9780199548798.001.0001/oxfordhb9780199548798.

Earnshaw RA and Vince JA. 1995. Multimedia Systems and Applications. London: Academic Press.

Grace Kite. 2012. The Impact of Information Technology Outsourcing on Productivity and Output: New Evidence from India Volume 1. Pages 239-48.

ST Nandasara. 2009. Information Communication Technology – Grade 11. Educational Publications Department, Ministry of Education, Sri Lanka https://www.researchgate.net.

Thatchinamoorthi J and C Meenambigai. 2018. Textbook of Extension Communication & Information Technology. ISBN-10: 9788183214681 ISBN-13: 978-8183214681.

Vanaja and Rajasekar. 2016. Information & Communication Technology (ICT) In Education. First Edition, Neelkamal ISBN- 0: 8183165192 ISBN-13: 978-8183165198. https://www.india.gov.in/topics/communication/information-and-technology

Smart Villages Through Information Technology – Need of Emerging India https://www.researchgate.net/publication

For Product Safety Concerns and Information please contact our
EU representative GPSR@taylorandfrancis.com Taylor & Francis
Verlag GmbH, Kaufingerstraße 24, 80331 München, Germany